Dr Paul Parsons is a regular contributor to *Nature*, *New Scientist* and the *Daily Telegraph*. He was formerly editor of the BBC's award-winning science and technology magazine *Focus*. *The Science of Doctor Who* (Icon Books) and was longlisted for the Royal Society Prize for Science Books. His last book was *Science 1001*, published by Quercus.

HOW TO DESTROY THE UNIVERSE

And 34 Other Really Interesting Uses of PHYSICS

PAUL PARSONS

Quercus

This paperback edition published in 2012 by

Quercus
55 Baker Street
7th Floor, South Block
London
W1U 8EW

Hardback edition published by Quercus in 2011

A CIP catalogue record for this book is available from the British Library

PB ISBN 978 0 85738 837 7
EBOOK ISBN 978 0 85738 461 4

10 9 8 7 6 5 4 3 2 1

Illustrations by Guy Harvey and Nathan Martin

Typeset by Ellipsis Digital Limited, Glasgow

Printed and bound in Great Britain by Clays Ltd, St Ives plc

CONTENTS

INTRODUCTION

Why is it that when you read about physics in popular books, it's always about accelerating subatomic particles to near the speed of light in an attempt to unlock the ultimate secrets of the Universe, and yet when you study it at school all you end up doing is measuring the temperature of some ice in a bucket?

Perhaps that's an exaggeration but it's not really a surprise that for all too many people, physics lessons were boring. Tediously, mind-achingly, duller than defrosting the fridge on a rainy Sunday, boring.

When I was at school I had two physics teachers. One, Mr H, spoke with a lisp and walked like the soles of his shoes were made of Zectron – that super-springy stuff they used to make balls that, if you lobbed them at the ground hard enough, could bounce right over your house. Despite his comical comportment he was, sadly, a droning bore. Albert Einstein once remarked how odd it is that an hour spent in the company of a pretty girl seems like a minute, while a minute with your hand on a hot stove seems like an hour. 'That's relativity,' he said. If only the great man could have come along to one of Mr H's classes, he could have witnessed time actually appear to run backwards. I developed a deep loathing for the sections of the syllabus that Mr H inflicted upon us – which, included thermodynamics (the science of temperature, ice and, yes, buckets).

My other physics teacher – Miss M – was four foot ten and, so the story goes, had the power to make the school bully blubber without even raising her voice. Neither I nor any of my friends sympathized with school bullies but, nevertheless, we all regarded Miss M as quite terrifying and definitely not one to be aggravated. Homework was delivered promptly. That said, she was also perhaps the best physics teacher in the world. The vagaries of radioactivity, wave theory, gravity, optics, and all that other stuff, suddenly became clearer than centrifuged Evian. Not only that, but I don't ever recall being bored. Scared, yes. Bored, definitely not.

Thanks to Miss M, a very mediocre secondary school physics student was able to go away to university and ended up completing a doctorate in cosmology. Yes, that was me. I say 'able to' but perhaps 'wanted to' was her biggest achievement. I started off with next to no interest in physics, education or having a career of any sort, and came out of school inspired, largely as a result of her efforts.

But why should it take such a good teacher to make physics interesting? Physics, I think it's safe to say, is the best of all the sciences. That's not just because it covers nuclear explosions, which are the biggest explosions we're able to make. Or because it deals with space, which is inherently cool. It's more because physics is the most fundamental of all the sciences. As the great Ernest Rutherford – the man who first split the atom – once declared, 'Physics is the only real science. The rest are just stamp collecting.'

I think what Rutherford meant is that physics underpins the fundamental behaviour of the Universe – from that, everything else follows. The interplay between the subatomic particles – in particular, electrons orbiting around atoms – is what determines the laws of chemistry. And biology is just the chemistry governing the strange set of chemical reactions we call life. Life is classified into families and species – but giving things names and maintaining lists is no more innovative than keeping stamps in an album ... But I digress.

This book is your very own Miss M. I hope it won't scare you quite as much as she scared me and my friends, but the aim in writing it was much the same as her goal in teaching us: to provide an interesting and accessible guide to the big ideas in physics. I don't mean just the usual interesting fare of relativity and subatomic particle physics, but also mechanics (the science of moving objects), electromagnetism (the science of electric and magnetic fields) and even thermodynamics (temperature, ice and buckets). Along the way, I've tried to include some history of the subject and to put it all in a real-world context so that it doesn't all seem like blue-sky science.

Of course, there's plenty of blue-sky science in here too – relativity and subatomic particle physics, along with antigravity, parallel universes, teleportation, time travel, immortality, invisibility and higher dimensions of space and time. You'll find out how to save the planet from energy shortages by mining the vacuum of empty space, engineer the Earth's climate to reverse the effects of global warming, and fend off killer asteroids like Bruce Willis and his vest. You'll learn essential survival skills such as how to live through a lightning strike, tough it out during an earthquake and fall into a black hole without being squashed into spaghetti. And you'll discover some plain old cool stuff like how to turn lead into gold, travel to the centre of the Earth, crack supposedly unbreakable codes and use physics to predict the stock market.

Look at it this way: I got a physics education; you're getting the keys to world domination. Is that a good deal? This one's for you, Miss M – we salute you!

CHAPTER 1

How to build the ultimate rollercoaster

- Gravitational energy
- Launch catapult
- G-forces
- Centripetal force
- Mind the gap

Being accelerated from zero to 100 km/h (60 mph) in a little over a second, turned upside-down, spun round at over five times Earth's gravity and then dropped 100 m (330 ft) might not be everyone's cup of tea. But for rollercoaster thrill junkies it's their idea of heaven. The ultimate rollercoaster ride is a delicate balancing act between safety and being scared witless.

Gravitational energy

After an age spent queuing you finally climb aboard, buckle in and wait anxiously for the off. You've never done this before and aren't quite sure what to expect, although the green-faced individuals you've just watched stumble from the ride give you a fairly good idea. Amid fleeting concerns for your wellbeing, the controller's voice crackles over the tannoy: 'Go, go, go!' The car lurches forwards and starts to accelerate. Most rollercoaster cars do not have their own internal power source. In fact, they are not propelled at all for most of the duration of their journey. Instead, they are hauled to the top of a high peak and then released. It is the speed the cars gain during this initial drop that provides the energy needed to carry them around the rest of the track. The rollercoaster really does 'coast' the majority of the way. That this is possible at all comes down to a central principle of physics known as the 'conservation of energy'. It says that when you add up the amount of all the different forms of energy locked away in a physical system you get a number – the total energy of the system – that must remain constant with time. Energy in the system is allowed to change from one type into another, but the sum total must always be the same.

In a rollercoaster, the principal kinds of energy are kinetic energy, which is the energy associated with the motion of the rollercoaster cars, and 'gravitational potential energy' – the energy the cars possess because of their height in Earth's gravitational field, which can be thought of as rather like the energy stored in a stretched spring. At the peak marking the start of the ride, the rollercoaster's speed and kinetic energy are both zero. All of its energy is in the form of gravitational potential energy. When it is released and begins to fall, it steadily gains speed, converting gravitational energy into kinetic energy as it descends – and back again as it climbs. In reality, this conversion is not perfect, as some energy will be lost due to friction between the wheels and the track and between the wheels and other moving parts of the rollercoaster. Friction is caused when the microscopic lumps and bumps on two surfaces chafe against one another as the surfaces rub together. There is also friction between the rollercoaster and the air. The lost energy is not destroyed but is carried away in the form of heat and sound. The loss of energy to friction means that all the peaks on a rollercoaster course must become progressively lower than the starting point. If any of the peaks were the same height (or higher), the rollercoaster would not have enough energy to clear them. Instead, it would roll back down into the last valley, oscillating back and forth in the dip as friction gradually carried the rest of its energy away, ultimately bringing it to a stop. While putting the dampers on most of the ride, friction is essential if you ever intend to stop and get off. It's how the brakes work on most rollercoasters – by applying friction pads to the rotating axles to deliberately turn the rollercoaster's kinetic energy into heat as quickly as possible.

Conservation of energy is a concept that applies right across the whole of physics. It is an important principle in wave theory, thermodynamics, quantum mechanics and relativity. In 1918, German physicist Emmy

Noether proved that the conservation of energy is a direct consequence of the laws of physics being 'time invariant': meaning that if I drop a stone out of my bedroom window today, then it will fall to the ground in exactly the same way if I repeat the experiment tomorrow.

Launch catapult

Of course, not every rollercoaster relies on gravity. Some of the newer designs incorporate launchers to provide the initial boost to gets things moving. These employ mechanical catapults, electromagnets or hydraulic systems that make use of compressed liquid to give the cars a kick down the track. For example, the hydraulic launcher used on the Stealth rollercoaster at Thorpe Park, England, accelerates the cars from 0 to 130 km/h (80 mph) in just two seconds. That's an average acceleration of 18 m/s (60 ft/s) every second, roughly twice the rate you would accelerate by if falling freely under gravity. Physicists call this an acceleration of 2G. It creates a force pushing you back into your seat that is twice as powerful as the gravitational force on your buttocks as you sit reading this. G-forces such as this are an essential part of any rollercoaster experience. You feel them when the rollercoaster is accelerating forwards (in the case of launched rollercoasters), accelerating backwards (i.e. during braking – this normally only happens at the end of the ride) or changing direction.

G-forces

Changes in direction can take place in the vertical plane (passing over a crest or through a dip) or in the horizontal plane (turning a corner). The G-forces you experience in each case will vary according to what it's safe for the human body to experience. The highest permissible forces are those pushing you into your seat at the bottom of a dip. These can briefly reach up to 6G. By comparison, astronauts on the Space Shuttle rarely experience more than 3G. (Although, admittedly, astronauts must endure high G-forces for many minutes during the trip into orbit, whereas on a rollercoaster they last just a split second.) The opposite forces, which lift you out of your seat as you pass over a peak, are typically much lower, at around 2G. The weakest forces are those experienced on rounding a horizontal bend. These should not exceed 1.8G, owing to the weakness of the muscles in the side of the human neck. Most rollercoasters try to ease these lateral forces by banking the track on bends so that some of the cornering force is transmitted down through the body and into your seat rather than pulling sideways on the neck.

The forces you feel when you go round corners are all

down to Newton's laws of motion. These are three laws of physics that English physicist and mathematician Isaac Newton first published in his book *Mathematical Principles of Natural Philosophy* in 1687. The first law of motion says that an object will either remain stationary or carry on moving in a straight line at constant speed unless a force acts on it. This is sometimes also known as the law of 'inertia'. It means that a rollercoaster on a straight and level track will carry on moving forever (assuming there's no friction). If the track turns, however, the rollercoaster turns with it. The passengers – which Newton's laws apply equally well to – have their own inertia and their own natural

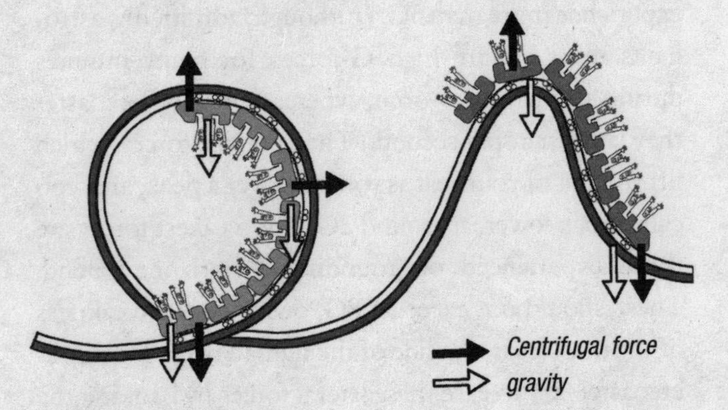

At the bottom of a loop centrifugal force and gravity both push you into your seat. At the top they work in opposite directions, so if the centrifugal force is strong enough it can overcome gravity and hold you in your seat.

If the centrifugal force exceeds gravity at a crest in the track it can produce negative G-forces, lifting passengers up out of their seats.

tendency to want to keep moving in a straight line. But instead they feel a force exerted on them by the side of the rollercoaster car as it turns.

Newton's second law of motion explains how the force makes the passengers turn the corner. It draws a distinction between forces and accelerations, and asserts that a force acting on an object causes the object to accelerate in the same direction as the force. If I push a toy car on a tabletop then I exert a force on the car, which makes it accelerate. Similarly, the passengers on a rollercoaster feel the force exerted on them by the car as it turns and as a result of it they are accelerated in a sideways direction.

Centripetal force

Sideways acceleration is also what enables a roller-coaster to loop-the-loop without you falling out of your seat. (All rollercoasters have restraints to hold you in, but in all but the slowest loop-the-loops these are unnecessary.) Here, the acceleration acts at right angles to the track, towards the centre of the loop, making the rollercoaster and the passengers move in a circle. At the top of the loop, where you are in the most danger of falling out of your seat, the acceleration pushes the seat into your bottom faster than gravity can pull your bottom out of the seat. As a result you stick to the seat. It's a similar effect that makes your washing stick to the sides of the spin dryer. Physicists refer to the force that causes this acceleration as 'centripetal force'. The strength of the centripetal force is determined by the radius of the loop and the speed at which the roller-coaster whizzes round it. The speed is lowest right at the top of the loop, but this is where the force needs to be strongest to stop you falling out. That's why the loops on some rollercoasters aren't circular but tear-drop shaped, with a section of tight curvature at the very top to give maximum centripetal force where it is most needed.

Although physicists prefer to talk in terms of centripetal force, most people are more familiar with 'centrifugal force' – a force acting in the opposite direction that seems to be pushing them down into the floor of the rollercoaster as it loops. Centrifugal force is a consequence of Newton's third and final law of motion, which states that for every action (that is, every force) there is an equal and opposite reaction (a force pushing in the opposite direction). So, for example, when I sit on a chair, the chair pushes back to support my weight and stop me crashing into the floor. You can also think of centrifugal force in terms of inertia – each passenger's inertia makes them want to keep moving forwards in a straight line at a tangent to the loop, in keeping with Newton's first law. As the rollercoaster car turns inwards, following the path of the loop, this inertia pushes the passengers down into the floor.

Considering the centrifugal force also makes it slightly easier to visualize the physics of looping the loop. At the bottom of the loop, both gravity and centrifugal force act in the same direction, making passengers feel extremely heavy in their seats. But at the top, the two forces practically cancel one another out, making the passengers feel almost weightless. It's up to the engineers designing the ride to make sure the centrifugal force at this point is just bigger than the force of gravity to keep people in their seats. Going over a crest in the track, passengers experience the opposite effect to looping the loop. It's rather like being on the outside of the spin dryer – the rollercoaster car drops away from under you faster than gravity can carry you after it, and you rise up out of your seat. Many rollercoaster junkies argue that these 'negative G-force' moments are some of the best parts of the entire ride.

Mind the gap

Suddenly you lurch forwards. The brakes are on and the ride is over almost as quickly as it began. As you disembark you try not to look too dishevelled in front of the people queuing up for their turn. But in reality, your internal organs feel like they've been through a food mixer, your head is pounding and you could swear you have bruised ribs from strapping yourself in too tightly. You vow to have another go before the day is out.

CHAPTER 2

How to predict the weather

- Weather watching
- How to read a weather map
- Predicting the weather
- Number crunching
- Climate modelling
- Chaos theory
- Strange attractors
- Super crunchers

On the night of 15 October 1987, the worst storm in 284 years tore across the south of England, battering homes and property and causing damage totalling £2 billion. Winds reached hurricane force and downed an estimated 15 million trees. And yet, just 24 hours before it struck, weathermen were laughing off suggestions that we might be in for a rough night. They predicted that the storm would fail to make landfall, and would bluster harmlessly up the English Channel. Way-off weather forecasting seems an all-too-common occurrence. But why is it so hard? And what can be done to improve it?

Weather watching

Human beings are obsessed with the weather. It dominates our small talk, stops us getting to work in the winter and regularly ruins public holidays in the summer. Hardly surprising then that our best brains have been trying to distinguish the makings of a balmy Sunday from those of a wet weekend for thousands of years.

In 1835, US scientist Joseph Henry used the newly invented long-distance electronic telegraph to set up a network of weather-monitoring stations across the United States, the readings from which were wired instantaneously to a central office at the Smithsonian Institution in Washington DC. Weather-monitoring stations use a variety of instruments to gather data such as temperature, air pressure, wind speed, humidity and rainfall. Today, the findings of ground stations are supplemented by ships, together with a host of eyes in the sky such as weather balloons, aircraft and satellites, which scan the state of the planet's atmosphere from all angles to get a handle on what the weather is doing now – and what it's going to do next.

How to read a weather map

Sometimes it is easy to draw up a basic picture of how the weather is going to behave. For example, if a ground station in Florida is registering low pressure and a ship off the coast in the Atlantic is reading high pressure, it's a good bet that Florida is due for strong winds as air rushes from the area of high pressure to the low. (Over larger scales winds are deflected by the Coriolis effect, caused by the planet's rotation – see *How to stop a hurricane.*)

Lines of constant pressure on a weather map are called isobars. They can be thought of as rather like contour lines on the 3D landscape you get by graphing the pressure at every point on Earth's surface. Pressure differences can sometimes be predicted from thermal effects. Hot air rises and the updrafts act to lower the pressure over warm regions, while cool downdrafts tend to create regions of high pressure. Warm updrafts carry with them moisture that forms clouds as it condenses at high altitude. Temperature differences sometimes appear on weather maps as warm fronts, denoted by a line of red semicircles, and cold fronts, marked out by blue triangles. The arrival of a cold front can cause rainfall – or 'precipitation', as meteorologists like to call it. Warm, moist air rises up above the advancing cold front where it condenses into clouds and then drops back to the ground as rain. When conditions are exceptionally cold the water can fall instead as snow or hail.

Predicting the weather

This broad-brush analysis is fairly straightforward, and allows forecasters to provide the public with a very general impression of the weather along the lines of 'tomorrow's going to be windy'. But what if we need more details – such as how fast the gusts will be in each area, what time of day they'll be at their worst, or indeed whether hurricane-force winds will plough up the English Channel or veer inland to wreak havoc? Predicting the weather in this much detail means solving the mathematical equations that govern the physics of the planet's atmosphere. These equations are fiendishly complicated, coupling together processes such as the fluid dynamics of the air and oceans, heat transfer, atmospheric chemistry and the physics describing the radiation arriving from the sun. In fact, they are so abstruse they're nigh on impossible to solve – at least by the conventional methods most of us used to solve equations in maths classes at school. Worse still, the equations are highly non-linear, meaning that small variations in the input variables can bring about wholesale shifts in the outputs, which makes it hard to even solve them approximately.

Number crunching

Physicists attack mathematical problems such as this using the only option left at their disposal: brute force. Or in other words, solving the equations 'numerically'.

This works by shoving best-guess numbers into the formulae and then tweaking their values by trial and error until the equations all balance up. The first person to suggest doing this for the weather was the British physicist and mathematician Lewis Fry Richardson. In 1922, he published a book called *Weather Prediction by Numerical Process*. In it, he imagined a vast hall filled with 'human computers': people armed with pen and paper all busily grinding out numerical solutions to the equations describing the weather. A central 'conductor' would collate their results and then issue them with new instructions. There was just one snag. Richardson calculated that keeping up with the world's weather in real time would require 64,000 of these mathematical drones – equivalent to the entire population of Palo Alto, California. It seemed the only way to realize Richardson's vision was to come up with a machine that could carry out the calculations automatically. And so it was that numerical weather prediction was put on hold for 20 years, pending the invention of the electronic computer.

Climate modelling

The first computer-based weather simulation was run in 1950 on a computer called ENIAC (Electronic Numerical Integrator and Computer) at the US Army Ballistic Research Laboratory in Maryland, where it had initially been used for working out artillery-shell trajectories. ENIAC's early weather models used an extremely simplified picture of the atmosphere, where the air pressure at any point is determined simply by the density. Gradually meteorologists built more sophistication into their models to account for the processes of heating and atmospheric circulation that generate our complex real-world weather phenomena.

Computer weather models are set up by dividing the atmosphere into a three-dimensional grid. British mathematician Ian Stewart, in his book *Does God Play Dice?*, likens it to a 3D chess board. The weather at each precise moment in time is determined by assigning each cube in the grid a set of parameters defining the temperature, pressure, humidity and so on within that cube. These numbers can be thought of as rather like the chess pieces. The computer then evolves the board forward according to the rules of the game, encoded in the physics equations describing the weather. The results amount to moving the pieces around on the board rather like moves in a game.

In each cube, the computer takes the values of all the weather parameters and crunches them through the equations to work out the rate of change of each parameter at that instant in time. The rate of change allows all of the parameters to be evolved forward by a short interval, known as the 'time step'. Now the new values for all the parameters can be fed back into the computer again and used to work out a new set of rates of change, which can then be used to evolve the whole system forward by the next time step, and so on. The process repeats iteratively until enough time steps have been accumulated to reach the point in the future for which the forecast is needed. For a model of global weather systems, the time steps might be ten minutes or so, but for simulations of the weather over small regions they can be as small as a few seconds. After each time step, the parameter values in each cell are meshed together to ensure continuity. The result is a model of Earth's weather that can be advanced as far into the future as needed.

Chaos theory

However, the model cannot just advance into the future. Something was still missing. The predictions of the computer weather models were still only good for a few days, after which time, they became hopelessly inaccurate. The reason why was uncovered in the 1960s by the US mathematician Edward Lorenz. What he found would revolutionize not just how we think about the weather, but pretty much the whole of maths and physics.

In 1963, Lorenz carried out a detailed study of the equations describing a key element to how the weather behaves: convection. This is the process that makes hot air rise and cold air sink. The same process happens in a pan of cold water that's heated from below on a stove. Even this small subset of weather maths was too difficult to solve on paper, so Lorenz put the equations on a computer. But when he did this he found something curious. If he stopped his simulation halfway through and wrote down the values of all the parameters, and then fed these back in manually to finish the simulation off, he got an answer wildly different from what he got by just letting the simulation carry on running in the first place. Lorenz eventually isolated the problem. Although the computer's memory was storing the numbers to an accuracy of six decimal places, it was only displaying its results to three decimal places. So, for example, if a number in the memory was 0.876351, the computer would only display 0.876. When Lorenz fed this truncated number back in, the loss of accuracy brought about by sacrificing those last three digits was skewing his results. So sensitive are the equations of convection to the initial conditions of the system that changing these conditions by just a few hundredths of a per cent was bringing about wildly different behaviour. Lorenz had discovered a phenomenon known as 'chaos': extreme sensitivity of a system to its initial state, meaning that tiny differences in that initial state become magnified over time. The main reason why forecasting the weather tomorrow is so difficult is because we cannot measure the weather today accurately enough. Lorenz even coined a term to describe the phenomenon – the 'butterfly effect', the idea that the tiny perturbations caused one day by a butterfly beating its wings could be amplified over time to create dramatic shifts in the weather days down the line.

Strange attractors

Today, chaos is known to crop up in all kinds of physical systems – including quantum mechanics, relativity, astrophysics and economics. Mathematicians spot the presence of chaos by drawing a diagram called a 'phase portrait', which shows how the system evolves with time. They look for areas of the phase portrait called 'attractors', to which the system's behaviour converges. Non-chaotic systems have simple, well-defined attractors. For instance, the phase portrait of a swinging pendulum is just a plot of the pendulum bob's position against its speed, and the attractor takes the form of a circle.

Chaotic systems have attractors with bizarre, convoluted forms known as 'fractals' – disjointed shapes that appear the same no matter how closely you zoom in on them. The simplest fractal is made by removing the middle third from a straight line and then repeating the process ad infinitum on the remaining segments. Edward Lorenz found that the attractor in the phase portrait of convection was indeed a fractal – a kind of

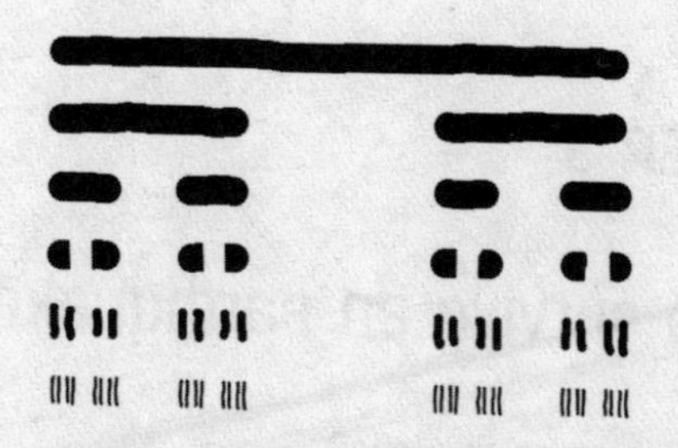

The simplest fractal is obtained by removing the middle third from a straight line and repeating the process.

distorted figure 8, which has since become known as the 'Lorenz attractor'.

Super crunchers

Improved computing power is now enabling the future evolution of chaotic systems to be predicted more reliably by storing the system parameters to a greater number of decimal places. The most powerful scientific computer is a modified Cray XT5, known as Jaguar, at the National Center for Computational Science in Tennessee. It has the same number-crunching capacity as about 10,000 desktop PCs. In truth, it's unlikely the weathermen will ever be able to tell us with 100 per cent certainty whether it's going to be sunny at the weekend. But disastrous misforecasts such as those that were issued prior to the Great Storm of '87 should at least become a thing of the past. Or so they tell us.

CHAPTER 3

How to survive an earthquake

- What is an earthquake?
- The magnitude scale
- Tsunamis
- Quake-proof buildings
- Mass dampers
- Earthquake prediction

Earthquakes are one of the most destructive forces in the natural world, equivalent in power to an atomic bomb. The quake that struck Haiti in 2010 killed over 200,000 people, and as cities in earthquake zones grow larger, it is becoming increasingly likely that a future quake could claim not thousands but millions of lives. Or is it? Are new technologies to mitigate the effects of earthquakes, ranging from giant pendulums inside skyscrapers to rubber feet under buildings, finally about to tame this awesome force of nature?

What is an earthquake?

Earthquakes occur when the tectonic plates that make up Earth's crust grate and grind against one another as they move. Tectonic plates are vast interlocking slabs of rock that float on the liquid layers of molten metal and rock that lie below them. As these liquids roll and froth, stirred up by the heat of the planet's interior, they drag on the plates above, pulling them this way and that. There are seven major tectonic plates – African, Antarctic, Eurasian, Indo-Australian, North American, Pacific and South American – and very many smaller ones. The boundaries where two plates meet are known as 'fault lines' and they come in a variety of different forms, depending on the relative motion of the two plates.

When the two plates are slipping past one another horizontally, the boundary is referred to by geologists as a 'transform fault'. As the plates jostle together, friction at the fault prevents them from slipping by smoothly. Instead they move in a jerking, juddering motion known as 'stick-slip'. First, the rock at the fault sticks because of friction. It deforms as the plates move, as if it were made of rubber. Over time the stress on the fault increases until eventually friction is overcome and the plates quickly slip past each other as the rock suddenly snaps back into shape.

An earthquake results when millions of tonnes of rock all rebounding in this way unleashes a violent mechanical wave that spreads out through the land, a bit like the ripple on the surface of a pond when you've

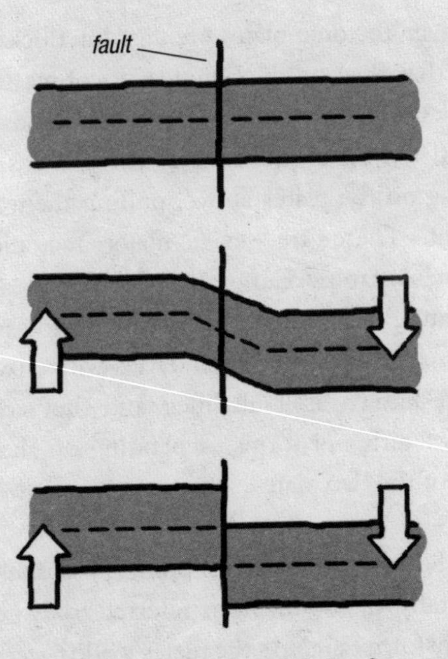

This is the view from above a geological fault line. Over many years, movement of tectonic plates deforms the landscape at the fault. When the build-up of elastic energy in the rock becomes great enough, it suddenly slips. This is an earthquake.

dropped a rather large stone in it. This wave, called a 'seismic wave', can have the power to bring down bridges and buildings, cause landslides and induce 'soil liquefaction' – where agitated soil assumes a liquid-like consistency, into which buildings and other structures can sink. Transform faults can spawn some truly destructive earthquakes, including the 1906 quake that devastated San Francisco, a city that lies next to the San Andreas Fault at the boundary between the Pacific and North American plates.

The magnitude scale

Seismic waves generated during an earthquake come in two different forms, called *P* waves and *S* waves. *P* waves are compression waves, rather like the waves you get on a stretched spring. The disturbance caused by *P* waves is parallel to their direction of motion. *S* waves, on the other hand, are more like water waves, where the disturbance is at right angles to the wave's motion, creating an S-shaped pattern of peaks and troughs as the wave passes. *P* waves travel roughly 1.7 times faster than *S* waves and scientists can use this fact to determine the distance to the earthquake's source, called the 'hypocentre'. Roughly speaking, eight times the time gap in seconds between the arrival of *P* waves and *S* waves gives the distance to the hypocentre in kilometres. By triangulating measurements made at a number of observing stations, the location of the hypocentre can be pinpointed. Most quakes happen within a few tens of kilometres of the surface, but the deepest ones can be located hundreds of kilometres down. The point on Earth's surface directly above the hypocentre is known as the 'epicentre'.

Seismologists gauge the power of an earthquake by taking its 'moment magnitude', which is a measure of the amount of energy the earthquake releases. This is an updated version of the Richter magnitude scale, first put forward by US physicist Charles Richter in 1935. Each increment in the scale corresponds to an increase in the energy of the quake by a factor of $10^{1.5}$ (about 31.6). In other words, an earthquake with a moment magnitude of 6 is 1,000 (31.6^2) times more powerful than a magnitude-4 quake. The 1906 San Francisco quake had a moment magnitude of 7.8, while Haiti in 2010 was magnitude 7. The most powerful earthquake on record, in Chile in 1960, measured a collossal 9.5. By comparison, the largest nuclear bomb ever detonated, the Russian Tsar Bomba in 1961, gave out energy equivalent to a magnitude-8 quake.

Tsunamis

Earthquakes don't just happen on land. In addition to transform faults, the two other kinds of fault boundary separating two tectonic plates are known as 'divergent' and 'convergent'. Here, the plates are either moving apart or slipping under one another, respectively. Divergent faults are normally associated with what are known as seafloor spreading sites, where new crust is being created at the bottom of the ocean. But far more lethal are the convergent faults, also normally found on the seafloor, where existing crust is sinking down into the planet's interior in a process called subduction.

Just as earthquakes at transform faults arise because of friction between the plates, so plates that are subducting undergo the same 'stick-slip' behaviour – as a large mass of rock suddenly springs back into shape having been deformed by the force of the moving plates. When this happens under water some of the energy of the quake is transmitted to the water, forming a tsunami, a giant wave, that rushes inland. These are known as 'thrust' earthquakes. The density of water (a single cubic metre weighs a tonne) makes them especially destructive. In 2004, a thrust earthquake off the coast of Indonesia measuring 9.2 on the moment magnitude scale (making it the second most powerful earthquake on record) threw up a tsunami that swept ashore killing 230,000 people. There are fears that a similar disaster could be waiting to happen off the coast of California, where the Juan de Fuca plate is subducting under the North American plate.

Quake-proof buildings

What can we do to protect ourselves in the face of such seemingly overwhelming might? The Inca civilization in Peru had a pretty good idea, 600 years ago. Many of their buildings, such as the complex at Machu Picchu, are still standing today despite being constructed in an area of extreme seismic activity. The Incas realized that making a building earthquake-proof isn't necessarily the same as making it stronger. The structures that survive today were built using a dry-stoning technique, where blocks of stone were stacked together with no mortar between them. The stones were so precisely cut that, so the story goes, you couldn't even shimmy a blade of grass between them. But when an earthquake struck, the lack of any mortar gave the buildings the flexibility to move and sway with the tremor, instead of crumbling under its force.

In the cities of the modern world, construction without mortar or other forms of fixing simply isn't an option. However, architects have managed to apply the Incas' logic elsewhere – in a building's foundations. The technique is called 'base isolation'. The building's superstructure (the part that's above ground) is coupled to its substructure (the foundations) using supports that are rigid under normal conditions but in the event of an earthquake become flexible, so that the vibrations in the substructure are not transmitted upwards where they could undermine superstructure. One example of such technology is known as a 'lead rubber bearing' – a support that sits under the building's superstructure and is made from rubber with a core made of the soft metal lead. The rubber makes the support flexible, while the lead serves as a 'damper' to stop the rubber getting too springy. The bearings can even be retrofitted into the foundations of existing buildings. Effectively, the buildings are being given a set of shock absorbers.

Mass dampers

Some modern skyscrapers incorporate giant pendulums in their upper levels. Known as tuned mass dampers, the pendulums are designed to swing inside the building at exactly the same frequency as, but in the opposite direction to, the swaying of the building. So as the building lurches to the left, the mass of the pendulum bob swings to the right to counterbalance it.

Tuned mass dampers are especially effective at combating a phenomenon called resonance, where vibrations at a structure's 'natural frequency' produce violent shaking that can lead to severe structural damage. To visualize how resonance works imagine a child playing on a swing. The swing makes one complete back-and-forth oscillation every two seconds. If the child's father, standing behind the swing, gives a push at exactly the same frequency – once every two seconds – each time the swing comes back towards him, the size of the oscillations will grow steadily bigger. Resonance is the reason why a truck with its engine idling will sometimes shake violently, but when the engine is revved to higher rpm the shaking subsides. Taipei 101, a 101-storey skyscraper in Taiwan, sports the largest tuned mass damper of any in the world. The bob of the damper pendulum weighs a mammoth 660 tonnes. The damper not only helps mitigate the threat from earthquakes but also serves to steady the building in high winds.

Earthquake prediction

Every year, one day in the third week of October, at 10.15 am, millions of Californians dive under tables, chairs and any other forms of cover they can find. Called the Great California ShakeOut, it's the world's largest earthquake drill. The annual drill is designed to help the sunshine state cope in the event of an unforeseen quake. Because that's what most quakes are: unforeseen. The science of earthquake prediction is, at best, shaky. It's rare for seismologists to predict accurately the date, time, location and magnitude of an earthquake. There is no way for them to gather data about rock movements deep underground. Normally, the best they can offer is probabilities. For example, after studying a particular fault for many years, using strain gauges to measure how the rock is stretching at the surface, they might be able to say that there's a 50 per cent chance of a 6+ magnitude quake, somewhere along the fault line, during the next 20 years. So what prognosis do the seismologists offer California? A study by the United States Geological Survey in 2008 concluded that the probability of an earthquake with a magnitude of 6.7 or higher striking the Greater Bay Area surrounding San Francisco, sometime in the next 30 years, is about 63 per cent – it's twice as likely to happen as not.

CHAPTER 4

How to stop a hurricane

- Hurricane Katrina
- The Coriolis effect
- Hurricane hotspots
- The Saffir–Simpson scale
- Project Stormfury
- Cool hurricanes

Hurricanes are the most devastating of all Earth's weather phenomena. They are fierce thunderstorms sometimes over 2,000 km (1,250 miles) across. Windspeed within a hurricane can reach 280 km/h (180 mph). And they can be accompanied by waves 10 m (33 ft) high that sweep ashore when the hurricane makes landfall. The average hurricane cranks out energy at a rate equivalent to a 10 megaton nuclear detonation every 20 minutes. Can we ever hope to be master over such a force? Some scientists think so.

Hurricane Katrina

On 23 August 2005, an innocuous-sounding weather system known as 'Tropical Depression 12' formed over the Bahamas. As it began edging its way towards the east coast of the United States, it grew in strength, reaching 'tropical storm' status early on 24 August. At this point it was also given a name: Katrina. The storm continued to gather momentum, reaching hurricane proportions just hours before crossing Florida and entering the Gulf of Mexico. By 28 August, it had strengthened to category 5, the most intense variety of hurricane. It made landfall in Louisiana at 6 am on 29 August with catastrophic consequences. At least 1,800 people were killed, and a further half a million were left homeless. However, Katrina is not the most devastating hurricane on record. That dubious honour goes to the Bhola cyclone that struck Pakistan and India in November 1970. The wall of water that accompanied it swept half a million people to their deaths.

Cyclones and hurricanes are essentially the same thing. The fundamental phenomenon is a cyclone – but cyclones cropping up in different parts of the world are given different names. A cyclone in the Atlantic Ocean, as was the case with Katrina, or in the eastern Pacific Ocean, is called a hurricane. One that arises in the western Pacific is known as a typhoon.

The Coriolis effect

Cyclones begin when warm ocean water causes moist air to rise high into the sky, as far as 15 km (9 miles) up.

At this altitude the air cools, releasing its heat, and causing the moisture to condense into rain clouds. The cool, dry air then falls back down to sea level where the cycle repeats. The physics responsible for this process is known as convection. It happens because gases expand as they are heated up. This lowers the gas's density, causing it to rise – in just the same way that an object with a density lower than that of water will float on the surface of the sea. Convection is also the reason why hot-air balloons are able to fly.

If only convection were involved, hurricanes wouldn't be much to write home about. But there's another process going on that stirs things up – quite literally. It's called the Coriolis effect, named after the 19th-century French scientist Gustav Coriolis, who first wrote down the mathematics describing it. It makes the air in Earth's northern hemisphere swirl in an anti-clockwise direction (as viewed from above), while air in the south swirls clockwise.

The Coriolis effect is caused by the planet's rotation. Imagine taking a series of horizontal slices through Earth from the North Pole down to the equator. As the planet turns all the slices rotate in lockstep, each slice completing one whole revolution per day. But the diameter of each slice gets bigger as you head south, so the actual straight-line speed of the slice's outer edge increases. For example, while Earth's surface at the latitude of New York (40.74°N) is travelling east at 1,260 km/h (783 mph), at the equator it's moving much quicker, at 1,670 km/h (1,038 mph). Someone in between the two, say on the island of Cuba (21.5°N) will be moving east at 1,554 km/h (966 mph). But here's the crucial thing. In this island dweller's own point of view, the equator is moving east relative to them at 116 km/h (72 mph), but New York actually appears to be going west, at 294 km/h (183 mph). The net result is to set up a turning effect that makes convection cycles, and other cloud masses in the northern hemisphere, swirl in an anti-clockwise direction. And this is why hurricanes and other cyclones spin. The Coriolis effect tends to produce a rising column of warm air that cools and spirals outwards at high altitude before it falls back to sea level, gets warmed once more by the ocean and then sucked back to the centre where it rises again. As the air rises and cools it releases its heat energy and this is what powers the hurricane.

Hurricane hotspots

Creating a sufficient thermal updraft to form a cyclone requires ocean temperatures of over 26°C (80°F). Generally the sea is only this warm within the tropics, which is why cyclones are sometimes known as 'tropical cyclones'. Cyclones can form in all the world's equatorial ocean basins. Each area has its own cyclone season, corresponding to the time of year when the difference between the temperature at sea level and at high altitude is greatest – driving the strongest convection currents. For the North Atlantic, this is June to November with most hurricanes occurring in August and September. In the southern Indian Ocean, the season runs from December until April. Once formed, a cyclone tends to migrate westwards, driven by the equatorial trade winds, which blow from east to west. Like hurricanes, the trade winds are caused by a combination of convection and the Coriolis effect. Warm air at the equator rises due to convection, cools and migrates outwards to latitudes of +/–30°, where it falls back to sea level. The low pressure that the convection causes then sucks this cooled air back down to the equator where the process repeats. If Earth did not rotate, this air would simply move in a straight line towards the equator, but the Coriolis effect changes that.

Again, it's rather like the situation in a hurricane. Here a patch of low pressure draws air currents radially inwards. But the Coriolis effect makes the currents form a swirling anti-clockwise vortex – in other words, each inward-moving current is deflected to the right. In fact, it's a general tendency of the Coriolis effect in the northern hemisphere to make air currents veer to the right; in the south it makes them veer to the left. And this is what makes the cool air currents heading towards the equator in the northern and southern hemispheres both veer west to create the trade winds at the equator. These winds blow cyclones in a westerly direction.

The Saffir–Simpson scale

Scientists classify the strength of hurricanes on what is known as the Saffir–Simpson scale, first put forward by US engineer Herbert Saffir and US meteorologist Bob Simpson in 1969. The weakest hurricanes are category 1, which have windspeeds between 119–153 km/h (74–95 mph); the strongest hurricanes are category 5, with winds exceeding 250 km/h (155 mph). The centre of a hurricane is a region of calm known as the 'eye', which is typically about 50 km (30 miles) across. The air pressure here is low because convection is at its strongest, sucking warm air away from sea level like a vacuum cleaner. The warm, rising air currents spiral up around the edge of the eye forming a thick bank of rapidly rotating cloud called the 'eyewall'. It's here that the strongest winds are found and rainfall is at its most torrential.

Project Stormfury

The strength of a hurricane diminishes rapidly once it reaches land, as its energy source – the warm ocean – disappears from under it. This means that while coastal regions are especially prone, moving 10–20 km (6–12 miles) inland is often enough to escape the worst effects. That works for saving lives, but what about property damage? After all, there's no way a whole city can be uprooted and moved to safer ground. Is there any way we could influence the physics underpinning a hurricane to alter its course or even stop it in its tracks? One of the earliest efforts to try to change the behaviour of hurricanes was the US Project Stormfury, which began in the 1960s. This was an effort to weaken hurricanes by stimulating rainfall within them using a technique called cloud seeding. The idea is for aircraft to fly above the hurricane and drop into it chemical particles with a crystal structure resembling that of ice, such as silver iodide. This encourages water vapour in the storm to cool and condense into clouds, which then fall to the ground as rain. It was believed this would cause the eyewall in a hurricane to get bigger. Just like the spinning ice skater pulls her arms in to go faster and stretches them out to slow down, so windspeed becomes lower as the eyewall gets larger. However, the results of Project Stormfury were inconclusive and it was abandoned.

Recently, US meteorologist Dr Ross Hoffman has carried out computer simulations showing the effect of heating hurricanes. He found that increasing the temperature at high altitude by just 2–3°C (4–5°F) can have a major effect on a hurricane's course. Heating the top of the hurricane decreases the vertical temperature gradient, weakening the convection currents that drive it. Hoffman suggests this heating might be achieved using a flotilla of satellites, which could bombard a hurricane with microwaves from above. Another proposal is to disperse soot particles in the hurricane's upper cloud layers. Soot, or indeed anything else black, absorbs heat. Clouds of it in a hurricane's upper decks would absorb sunlight, heating the clouds in much the same way as microwaves. Yet another option is to smother the hurricane in particles of Dyn-O-Gel, a polymer compound that can absorb as much as 1,500 times its own weight in water, soaking up the hurricane's heat-carrying moisture and robbing it of its energy source.

Cool hurricanes

Other scientists have proposed fighting hurricanes from the opposite direction – the ocean surface. One proposal is to lay down a slick of biodegradable oil over the sea to temporarily prevent warm water vapour from escaping. US entrepreneur Bill Gates has entered the fray, backing a plan involving barges that would sit in the path of a hurricane pumping cold water up from the ocean depths to cool the sea surface.

Many hurricane experts, however, are pouring cold water on these and other ideas. They argue that rapidly moving weather systems thousands of kilometres across are simply beyond the realms of human influence. Not only that, many worry about unforeseen consequences. Hurricanes, like other weather phenomena, are fiendishly difficult to forecast. Then again, given the potential for damage (over $100 billion in the case of Katrina), these may be technologies we cannot afford to ignore.

CHAPTER 5

How to deflect a killer asteroid

- Asteroid impacts
- The Tunguska impact
- Planetary defence
- The nuclear option
- Kinetic impact weapons
- The Yarkovsky effect
- Asteroid evolution

Sixty-five million years ago, the dinosaurs had a very bad day indeed. An asteroid 10 km (6 miles) across slammed into Earth and exploded with the force of 200,000 gigatons of TNT – 4 million times more destructive than the largest nuclear bomb. The blast gouged a crater 180 km (110 miles) across and unleashed global firestorms, quenched only by gargantuan waves that swept round the planet. Very little survived as the dinosaurs, along with many other species, became extinct. Asteroids bigger than 5 km (3 miles) in diameter hit Earth once every 10 million years. Will the arrival of the next one mean extinction for humanity? Not likely.

Asteroid impacts

Asteroids are chunks of rocky debris left over from the formation of the Solar System. As dust particles swirling around the young Sun 4.5 billion years ago randomly bumped into each other, they began to stick together like clods of soil. Eventually they became big enough for their gravity to draw in material that wasn't directly in their path, and they grew larger still. As these cosmic boulders collided, they continued to grow until a handful became so large that they formed into the planets and moons of the Solar System. Not that the collisions stopped. During the so-called 'late-heavy bombardment', about 4 billion years ago, the closest planets to the Sun were pelted with asteroids as the shrapnel left over from planet formation was pulled in by their gravity. On Earth, our atmosphere and weather have eroded all but the biggest craters – such as the Barringer crater in Arizona – but look no further than the pocked and scarred surface of the Moon for evidence of how unimaginably violent this phase of Earth's history really was.

The threat hasn't gone away. On 3 November 2008, at around 10.30 pm, a solitary asteroid whizzed past our planet at a distance of just 38,500 km (24,000 miles). It sounds like a wide birth, but in astronomical terms that's a mere hair's breadth – just a tenth of the distance from Earth to the Moon. And just a month before that, another space rock actually hit our planet with just a day's warning that it was coming, exploding in the sky over Sudan. Luckily the Sudan impact was a small one, just a few metres across, but the 3 November near-miss was much bigger. At 250 m (820 ft) in diameter and hurtling in from outer space at 20 km/s (12 mps) – nearly 60 times the speed of sound – the rock would have struck the ground (or exploded in the atmosphere as a result of heating caused as it compressed the air in front of it) with a force equivalent to a 500 megaton nuclear weapon – over 30,000 times the power of the Hiroshima bomb.

The Tunguska impact

Asteroid impacts threaten Earth with startling regularity. While asteroids of the size that wiped out the dinosaurs only strike once every 100 million years or so, smaller rocks come calling far more frequently. In 1908, an asteroid 45 m (150 ft) wide exploded in the skies over the Tunguska River in Siberia. The blast was big enough to flatten a modern city – indeed, had it landed on central London, everything within the M25 ring road would have been obliterated. Astronomers believe that Earth is pummelled by at least one Tunguska-size impact every few hundred years. After much lobbying from the international scientific community, the international political community agreed that action was necessary to combat this threat. The United Nations Working Group on Near-Earth

Objects held its first workshop in February 2009. Their goal is to coordinate the world's response to the detection of an asteroid on collision course with Earth.

Planetary defence

Most scientists believe that a lead time of several decades will be needed to do anything about a rock on a collision course with our planet. That means that as our first line of defence we must set up a cosmic early warning system. NASA has been given a congressional mandate to log 90 per cent of asteroids bigger than 140 m (460 ft) by 2020. The US space agency intends to detect and track hundreds of thousands of stony wanderers in the depths of space, monitoring them night after night so that their orbital trajectories through the Solar System can be established.

There's no way human astronomers working at their telescopes could achieve this, so most asteroid detection today is done by robot telescopes. Each night the telescope scans a predetermined area of sky and computer software compares its images to those taken on previous nights. Anything found to be moving from frame to frame, and which isn't a known asteroid or planet, is earmarked for further monitoring, and human astronomers are alerted.

Tracking a new asteroid over many nights tells the

astronomers how fast it's travelling. The mathematical laws of orbits then allow them to infer how far away it is and to work out its precise trajectory around the Sun. Comparing this to the orbit of Earth enables them to flag up any which look set to get uncomfortably close. But how close is close? Astronomers gauge the danger posed from near-Earth asteroids on what's called the Torino scale. It's rather like the Richter scale for earthquakes, and gives an indication of the danger posed by any particular asteroid. The name comes from a scientific conference held in Turin (Torino in Italian), Italy, in 1999, where the scale was first proposed. The scale ranges from zero up to 10. Zero corresponds to an asteroid that poses no risk at all, while 10 indicates a certain collision that's going to cause global devastation. The highest an asteroid has ranked on the Torino scale so far is asteroid 99942 Apophis, discovered in 2004, which scored a 4 – meriting concern but not alarm. Further observations later gave astronomers a better understanding of its orbit and they downgraded the risk to a 0.

The nuclear option

If and when the automated sky surveys do turn up a Torino-topping object, scientists have come up with a number of possible options. The long-standing favourite of Hollywood when it comes to dealing with calamitous rocks from space is nuclear weaponry. And yet experts agree that in most cases going nuclear is the worst course of action. The main problem is that atomic blasts are so violent that rather than deflecting an asteroid onto an orbit that misses Earth, the most likely outcome is to shatter it into a blizzard of smaller fragments that will still hit the planet. Any of these bigger than 30 m (100 ft) across (and it is likely there would be very many of these) will still be capable of penetrating the atmosphere and causing catastrophe.

Turning a rifle bullet into a shotgun blast like this also carries with it a hidden danger. Dotted around Earth are a number of points in space known as 'gravitational keyholes'. Although a rock passing through one of these keyholes will miss us on this pass, the planet's gravity will bat it onto a new path bound for collision one or more orbits down the line. With a spread of rocks heading towards the planet – as you'd get after a nuclear attack – it's likely that at least one of them will pass through one of these keyholes in space. Asteroid 99942 Apophis is due to make another close flyby of Earth in 2029 and, although it's unlikely to hit, there are concerns that it will pass through a keyhole, bringing it back to Earth in 2036.

Kinetic impact weapons

If nuclear is no good, then what are the other options? One method is to use a so-called kinetic impactor.

This is a solid mass with no explosive charge that slams into the asteroid and gives it a kick simply by virtue of its momentum, knocking it onto a new orbit. In 2005, NASA's Deep Impact mission did just that, firing a solid projectile at the nucleus of comet 9P/Tempel. The impact threw up a plume of material from the nucleus so that instruments on the probe could analyse the comet's composition. The mechanics of guiding a projectile to deflect a hazardous comet or asteroid are much the same. So much so, the European Space Agency is planning a mission to road test such a deflection technology. Called *Don Quijote*, it consists of a pair of spacecraft. One spacecraft will crash into a target asteroid at a speed of 36,000 km/h (22,000 mph) while the other measures how the rock's course is affected by the smash. ESA plans to launch the mission in 2011, but as of mid-2012 it is still only a study.

One of the deflection methods favoured by scientists is known as a gravity tractor. Heavy objects attract other heavy objects thanks to the force of gravity. In 2005, two former US astronauts – Edward Lu and Stanley Love – realized that this could be used to make a gravitational tow bar that could drag an Earth-crossing asteroid off its collision course with our planet. The basic idea is to have a spacecraft fly alongside the target asteroid. As gravity makes the asteroid and the spacecraft move together, the spacecraft fires its engines – and the asteroid follows. It's a neat idea, though you could argue that rather than going to all the trouble of sending a spacecraft to the asteroid and hovering near it for many years (because that's how long it would take to deflect an asteroid in this scheme), mission planners may as well just bolt a rocket motor directly to the asteroid. And that, too, is a possibility.

There is a more ingenious take on this idea, though – the mass driver. A rocket engine works by burning fuel in order to create a stream of high-speed exhaust gas that propels the spacecraft in the opposite direction. A mass driver does a similar job, but works by mechanically catapulting lumps of material in one direction to send the spacecraft – or, in this case, the asteroid – moving the other way. Rather like the kick from a gun firing bullets, each lump cast off produces a recoil, shoving the asteroid in the opposite direction. Over time, all the shoves add up to change its course. In practice, a mass driver would take the form of a solar- or nuclear-powered robot that sits on the asteroid's surface.

The Yarkovsky effect

Perhaps one of the strangest ideas for saving the world from killer asteroids derives from the work of 19th-century Russian engineer Ivan Yarkovsky. In general, asteroids spin as they travel through space. Yarkovsky showed that this spinning skews the way that heat is radiated from an asteroid's surface. This creates an acceleration on the rock that, over time, can alter its orbit around the Sun. As an asteroid rotates, it has a 'dawn hemisphere', the side where the surface is rotating from darkness into sunlight, and a 'dusk hemisphere' where the surface is rotating from sunlight back into darkness again. The dusk hemisphere is warmer (because its just been in bright sunlight) and so radiates more heat than the dawn hemisphere. Because the heat is carried away as photons of electromagnetic radiation, which pack momentum (see *How to harness starlight*), the radiation exerts a recoil on the asteroid that influences its orbit over time.

The Yarkovsky effect can be enhanced or diminished by changing the asteroid's colour, because different colours absorb and emit heat at different rates (in just the same way that black seat covers in a car get much hotter on a sunny day than white ones). This has led to the suggestion that one way to combat the threat posed by rocks from outer space is to send up a team of astronauts equipped with the world's largest paint rollers.

Asteroid evolution

Although generally regarded as detrimental to life, some scientists suggest that, ironically, the regular upheavals caused by rocks from space slamming into our planet presented formidable challenges to lifeforms emerging on the early Earth that stimulated their evolution – giving them resilience and the problem-solving ability needed to survive in this harsh environment. And asteroids may drive our own species to even greater accomplishments, giving us the motivation to migrate away from our planet and become a spacefaring civilization. As the dinosaurs discovered to their cost, Earth is a fragile cradle – one that human beings must ultimately leave if they are to continue to prosper in this often brutal universe.

CHAPTER 6

How to journey to the Earth's core

- Earth's anatomy
- The core
- Earth's deepest holes
- Reasons to dig
- Probe to the core

Human beings have explored some of the most far-flung reaches of the Solar System, but when it comes to investigating the innards of our own planet we've barely scratched the surface. Jules Verne's 1864 novel *Journey to the Centre of the Earth* told the tale of a group of explorers who venture down into the bowels of the planet. Verne's heroes discover a subterranean world populated by dinosaurs and prehistoric humans. But what really lies at the heart of our planet? Scientists think they might have the means to find out.

Earth's anatomy

Our planet is a 4.5-billion-year-old ball of rock and metal just over 12,700 km (7,900 miles) in diameter. If you could take a knife and cut Earth down the middle from pole to pole, you'd find within a layered structure rather like the inside of an onion. The outermost layer is known as the crust. It is made of various kinds of rock, and its thickness varies considerably. Beneath the oceans it can be as little as 5 km (3 miles) deep, whereas on land the thickness of the so-called continental crust can reach up to 40 km (25 miles). This is the primary reason why the continents are raised up out of the ocean. The crust accounts for roughly 1 per cent of the volume of the planet.

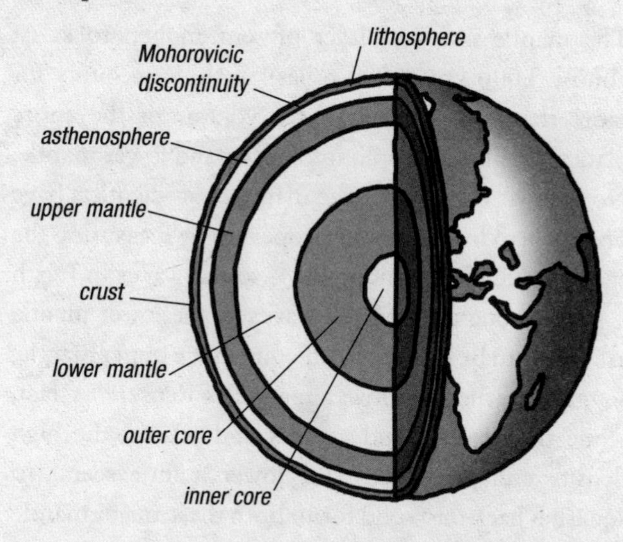

The terrestrial onion: what travellers inside Earth can expect to find.

Temperature increases markedly as you dig down into the crust, sometimes by as much as 2–3°C (4–5°F) for every 100 m (33 ft) down you go. And so the temperature in a mine 3 km (2 miles) below the surface can rise to nearly 100°C (212°F). This is mainly due to the compression effect caused by the weight of the overlying rock and soil – just as the compressed air in a bicycle pump feels hot to the touch. Radioactivity in rocks can also contribute. The rate of increase of temperature with depth slackens off in the lower regions of the crust, reaching a maximum of about 400°C (750°F) at the boundary with the next layer down, known as the mantle.

The mantle is a soft layer of semi-molten rock. At almost 3,000 km (1,900 miles) thick, it accounts for more than 80 per cent of the volume of the entire planet. It is broadly split into upper and lower mantle. No one has ever visited the mantle, but scientists have been able to investigate its properties by measuring the speed at which seismic waves – sound waves in Earth – pass through it. Seismic waves in the lower mantle are found to be faster than they are in the upper mantle, suggesting that the lower mantle is denser. In fact, while the upper mantle is soft and pliable, the high density and pressure in the lower mantle seems to squash it back into solid form. Both these major mantle layers are heavier than the overlying crust, which is formed from low-density rock that melted and then rose up to float at the surface like a cork in water.

The upper mantle divides into a number of sub-zones. The lithosphere is made up of the crust together with the stony outer shell of the upper mantle; the asthenosphere is a softer shell of mantle directly below this; and finally the Mohorovicic discontinuity, or 'Moho layer', is the boundary that separates the mantle from the crust. The upper mantle reaches down to a depth beneath the planet's surface of around 400 km (250 miles). Between its upper and lower zones is a layer known as the transition zone. From there, the lower mantle extends down to a depth of 2,900 km (1,800 miles), at which point the temperature has soared to a blistering 4,000°C (7,000°F). This is the beginning of Earth's core.

The core

The core is the densest, hottest part of Earth's interior. It's made mainly from the metals iron and nickel, and, like the mantle, it is divided into an inner and an outer layer. The outer core is mostly liquid and reaches down from the bottom of the mantle to 5,150 km (3,200 miles) beneath the planet's surface. The inner core is a ball of solid metal within this, roughly 2,440 km (1,500 miles) in diameter. The outer core is thought to be responsible for generating Earth's magnetic field. As electric currents in the conducting liquid metal are swished round by the planet's rotation, they create a magnetic field through the dynamo effect – exactly the same principle by which electrical generators work. Although the field is very weak – about one thousandth the strength of a fridge magnet – it's strong enough to bat away electrically charged cosmic ray particles that wander in from space. A single cosmic ray can contain as much energy as a fast tennis serve – all packed into a tiny subatomic particle. If this much energy crashes into a DNA molecule in your body it can cause mutations that could lead to cancer and death – indeed, life on Earth would have had a hard time getting going were it not for the core and its magnetism. We owe our very existence to it.

Some scientists have speculated, however, that Earth's magnetic field may fail us yet. Every few tens of thousands of years, the field undergoes a complete reversal – where north and south poles literally trade places. There has been concern that as the field reverses its strength could temporarily diminish, allowing harmful cosmic rays to reach the planet's surface. The next reversal is due to begin in the next few thousand years.

Earth's deepest holes

The deepest people have ventured below Earth's surface is 3.9 km (2.4 miles), down the TauTona gold mine in South Africa. Getting to this depth in the mine shaft's elevator takes an hour and the temperature of the rock face once you get there reaches 60°C (140°F). The only way human beings can function in this heat is by fitting the mine with a sophisticated cooling system that pumps refrigerated air into the tunnels. But humans have managed to go deeper than this, albeit indirectly – that is, by drilling down into the planet from the surface. The deepest borehole ever dug into Earth went down 12 km (7 miles). The hole was drilled by Soviet scientists working in the Kola Peninsula east of Finland, and penetrated about a third of Earth's crust at that point. It took 19 years to drill and was completed in 1989. The scientists had hoped to continue drilling deeper still, but at a depth of 12 km (7 miles) they found the temperature was rising above the threshold at which their drill could operate.

The Kola Borehole gave scientists access to rock from a period of Earth's history known as the Archaean Aeon, 2.7 billion years ago. Studying rock from such an ancient era has helped to refine ideas about how Earth was formed and how it evolved into the world we see today. It is also shedding light on modern-day climate change. In 2004, samples of seafloor rock that had been retrieved by drilling into the Lomonosov Ridge beneath the Arctic ice sheet revealed that 55 million years ago the region had been so warm that it had no sea ice – the North Pole was located amid a liquid water ocean. Studies of how the climate responded to such periods of warmth in the past may be the key to understanding the consequences of future global warming.

Reasons to dig

The wealth of knowledge that can be gained from exploring Earth's interior has led to a renewed interest in drilling deeper. An international collaboration of geologists known as the Integrated Ocean Drilling Program (IODP) is trying to punch a hole all the way through Earth's crust and down into the Moho layer, which marks the start of the semi-molten mantle underneath. Geologists have long been on a quest to reach the elusive Moho layer. In the 1950s, the US Project Mohole was proposed to drill down to the Moho to help work out how movements of the mantle influence the crust above.

Heat circulates in the mantle by rolling convection cycles – the same process that makes warm air rise and cool air sink. These churning currents of molten magma drag on the crust, causing it to move. This effect makes the crust heave and crack, leading to volcanic eruptions. It is directly responsible for the motion of Earth's tectonic plates, which is the cause of earthquakes (see *How to survive an earthquake*). But the details of exactly how this all happens are poorly understood. Project Mohole was cancelled before it ever got going. But now the IODP has picked up the baton, using two specially equipped drilling ships to try to penetrate the crust beneath the ocean – where it's thinnest. The goal is to open a window on the deep Earth that will give scientists fresh new insights into the physics responsible for some of the planet's most destructive outbursts.

Probe to the core

On a planet many thousands of kilometres across, boreholes just a few kilometres deep barely prick the skin. One physicist, however, has come up with a plan by which we could send a scientific probe all the way down to Earth's core. In a paper published in 2003 in the respected science journal *Nature*, US planetary scientist David Stevenson proposed using a multi-megaton nuclear bomb to open up a vast crack in the planet's crust. Into this enormous crater would be poured 100,000 tonnes of molten iron. It sounds like a lot but that's the same amount that's turned out by all of the world's foundries in the space of about a week. The sheer weight of the iron – it's about twice as dense as the rock in Earth's crust – would then make it sink, causing the crack in the ground to spread downwards. As the iron sinks the pressure within Earth closes the crack behind it.

Stevenson calculated that a grapefruit-sized probe thrown in with the iron would drop down to the core in about a week, where it would be able to measure details such as the precise temperature, pressure and chemical composition. The probe would transmit its findings back to the surface via seismic waves – the same sound waves that geologists use to gauge Earth's interior structure remotely. Stevenson's plan could be put into action for a total price tag of around $10 billion. This is much less than the total amount the United States spends on space exploration every year (NASA's total budget for 2010 was nearly $19 billion), and could help plug a massive gap in our knowledge – we know more about some faraway planets than we do about the ground beneath our very feet.

But where human explorers have often followed in the footsteps of their robot counterparts in space, it seems unlikely there will ever be any attempts to visit Earth's core in person. The temperature there rises to around 7,000°C (13,000°F) – hotter than the surface of the Sun. The pressure is over 3 million times the planet's atmospheric pressure. These conditions are far too extreme for the centre of Earth to be a haven for the extinct species Verne envisaged – or indeed any known life forms whatsoever.

CHAPTER 7

How to stop global warming

- The greenhouse effect
- Air pollution
- The tipping point
- Fixing a broken planet
- Sun blockers
- Terraforming

Earth is getting hotter. Estimates in 2009 suggest that if current trends continue global temperatures could rise by as much 5°C (9°F) this century, bringing droughts, extreme weather and sea-level rises of several metres that will threaten coastal cities around the world. Growing evidence suggests that the current acceleration in climate change is caused by the chemical by-products of our industrial civilization. Now some scientists think this same civilization might be able to use its technological know-how to undo the damage.

The greenhouse effect

The principal contributor to climate change is the greenhouse effect, where the atmosphere traps some of the heat that arrives from the Sun. It happens because the atmosphere is partially opaque to infrared radiation, which is the type of radiation by which most heat is transmitted. How radiation travels through a material is determined by the material's atomic and molecular structure. Atoms are made of a central, positively charged nucleus with electron particles orbiting around it. The electrons can each occupy one of a well-defined set of orbits around the nucleus, and each orbit has an energy associated with it. The difference between two orbits forms an energy gap. Radiation with energy equal to the size of the gap can be absorbed, causing the electron to jump from the lower-energy to the higher-energy orbit. Different atoms have their own characteristic set of energy levels enabling them to absorb radiation at particular energies. Similarly molecules – made from atoms bolted together – have their characteristic energy levels.

The chemicals in Earth's atmosphere absorb radiation at many different wavelengths. The primary absorbers of infrared are water vapour (H_2O) and carbon dioxide (CO_2). You might think that if the atmosphere is opaque to the Sun's heat then, if anything, the planet should be getting cooler. However, the Sun doesn't just bombard Earth with infrared, but radiation from right across the electromagnetic spectrum – including radio waves, visible light and ultraviolet. Much of this can pass through the atmosphere and makes its way down to the ground. Here, it is absorbed by the soil, rocks, oceans and buildings, which then re-emit it as heat. And it's this extra heat that gets trapped and warms up the planet.

Air pollution

The principal man-made gas contributing to the greenhouse effect on our planet is carbon dioxide (CO_2). Levels of atmospheric CO_2 are now believed to be at the highest they've ever been in the last 15 million years. Scientists know this from studies of ice cores gathered from deep beneath Earth's polar caps, and from marine sediments, which together serve as a kind of fossil record for the planet's atmospheric composition. Scientists have also managed to trace Earth's temperature back through time. This is possible because the thickness of ancient tree rings is linked to the length of each year's growing season, which is in turn coupled to how warm the climate is. Plotting the temperature over the last thousand years leads to a graph known as the 'hockey stick' because of its shape.

The hockey stick is a plot of temperature against time and it shows a flat line with a steep upturn around the late 19th century – as industrialization started worldwide. Even so, it wasn't until the 1970s that better understanding of the climate led scientists to realize that rising temperatures could have catastrophic results. If all of the ice on Earth melted, it would be enough to raise the sea by 70 m (230 ft). That scenario is unlikely, but even a rise of 1–2 m (3–6 ft)would be enough flood major cities – including London, New York and Tokyo. Meanwhile the damage to crops and freshwater supplies would threaten billions of people.

The tipping point

Coal, petrol and other fossil fuels give off copious amounts of CO_2 when they're burned. That, per se, isn't a problem. If you burn a big stack of wood it will also give off a lot of CO_2. The difference is that the CO_2 released when wood burns was already in the climate system – the tree soaked it all up before it was chopped down for fire wood, so what gets released is simply going back where it came from. The gas given off by burning fossil fuels, on the other hand, was previously locked away underground – and so this is new CO_2 that's being added to the environment. This is why efforts to develop 'biofuels', replacements for petrol that are derived from renewable plant materials, are a promising possibility. (Although there is concern about the amount of land taken up to farm biofuel crops, and the impact this in itself will have on the environment.) Few people doubt that we need to cut down our carbon emissions, though many scientists worry that it will be too little too late. Their concern is that the state of the environment may soon cross a so-called tipping point, beyond which it will be extremely difficult to reverse the climate's freefall into global meltdown. 'Tipping point' is a phrase used by mathematicians to describe a sudden, discontinuous hop from one state to another. For example, gradually increasing the weight on the end of a rope will ultimately lead to a tipping point as the rope reaches its breaking load – and once the rope has snapped there's no going back.

But there is a glimmer of hope in all this. After all, if the current global warming is our fault then we've already demonstrated our ability to influence the planet's climate in a big way. And if we can do that, can we influence the climate in the opposite direction and put things back how they were? Some researchers think so. They've been devising schemes to patch up the damage – a field of science known as 'geoengineering'.

Fixing a broken planet

The options for geoengineering break down broadly into two categories: soaking up carbon from the atmosphere and blocking solar radiation. The simplest way to soak up carbon from the atmosphere is to plant more trees. Trees – and all green-leaved plants – take in CO_2 in order to produce their own food. The process is called photosynthesis, a chemical reaction whereby CO_2 and water combine with sunlight to make energy-rich carbohydrate plus oxygen, which is emitted back to the atmosphere and which all animals need to breathe.

If you can't plant real trees, one idea is to build artificial ones. The 'trees' resemble giant fly swatters that sift CO_2 from the air that flows through them. They work by absorbing the gas into a solution of sodium hydroxide. This is then heated in a kiln, causing the CO_2 to be given off as steam, which can be captured and bottled in high-pressure tanks. The gas in the tanks is then compressed into a liquid, which is pumped underground, for example into the cavities left behind by disused oil wells: putting the CO_2 released by burning fossil fuels back where it came from.

A less reliable option that has been suggested is to add vast amounts of fertilizer to the oceans, such as iron or the nitrogen-rich compound urea. The idea is that the fertilizer will encourage the growth of phytoplankton, which feed on carbon during the course of their lives. When the plankton die, their bodies, along with all the carbon they've absorbed, sink to the ocean floor and ultimately get buried by sediments. The risk with this plan is that unbalancing the chemistry of the ocean ecosystem with the vast amounts of chemicals that would be needed could do more harm than good.

Sun blockers

Blocking light and heat from the Sun is another major plan. One proposed solution involves spraying sea water into the air. As the water rises into the atmosphere, it evaporates, leaving tiny particles of salt that stimulate clouds to condense around them, a process known as 'cloud seeding'. The clouds would reflect solar radiation away into space. A fleet of robotic ships would do the spraying. The main drawback with this scheme is that no one knows for sure that it will work. The role of clouds in climate models is notoriously uncertain: white clouds do indeed reflect solar radiation, but water vapour, like CO_2, is a greenhouse gas that will trap heat and contribute to global warming. Which of these two effects wins out is at present unclear.

Another idea is to belch millions of tons of sulphur particles into the atmosphere where they would blot out some of the Sun's light. This technology has already been tried and tested – though not by humans. In 1991, the volcano Mt Pinatubo in the Philippines erupted, spewing an estimated 20 million tonnes of sulphur dioxide skywards, reducing global temperatures by 0.5°C (1°F) for two years following the eruption. The major disadvantage with this plan is that sulphur in the air is what causes acid rain, thanks to a chemical reaction in which sulphur dioxide combines with oxygen and hydrogen to make sulphuric acid. Acid rain increases the acidity of lakes and oceans, killing fish, which in turn impacts upon species that feed on the fish, sending a ripple effect all the way up the food chain. A less-polluting yet slightly more ambitious alternative is to launch a fleet of reflective parasols into space. These would sit at the L1 Lagrange point between Earth and the Sun, where the gravitational fields of the two bodies to some extent cancel out. A space parasol placed on the line connecting Earth and the Sun feels the gravity of both bodies pulling it in opposite directions. Too close to Earth and it will fall towards the planet, too close to the Sun and it will fall that way instead – but right between the two is a point where it will stay put, orbiting the Sun in lockstep with Earth. This is the L1 point. There are another four Lagrange points, labelled L2–5, dotted around the Earth–Sun system.

But this plan won't be cheap. It's estimated the combined area of the sunshade would need to be about the size of Greenland, and that would mean lofting some 20 million tonnes of hardware into space. At present, the most inexpensive space launchers can do this at a cost of around $4,000 per kilogram of payload, meaning a total cost for the sunshade of $80 trillion, more than the GDP of the entire planet.

Terraforming

Some ambitious space engineers want to use geoengineering principles not just on Earth, but on other planets too. Called 'terraforming', this involves sculpting the temperature and atmospheric composition, and introducing surface water. The most obvious candidate for terraforming in our Solar System is our next-door neighbour Mars. Scientists have suggested that the Red Planet could be made more like Earth by introducing plants that have been genetically modified to exist in its harsh climate. Through photosynthesis, these would gradually introduce oxygen into Mars's atmosphere – which is currently 95 per cent CO_2.

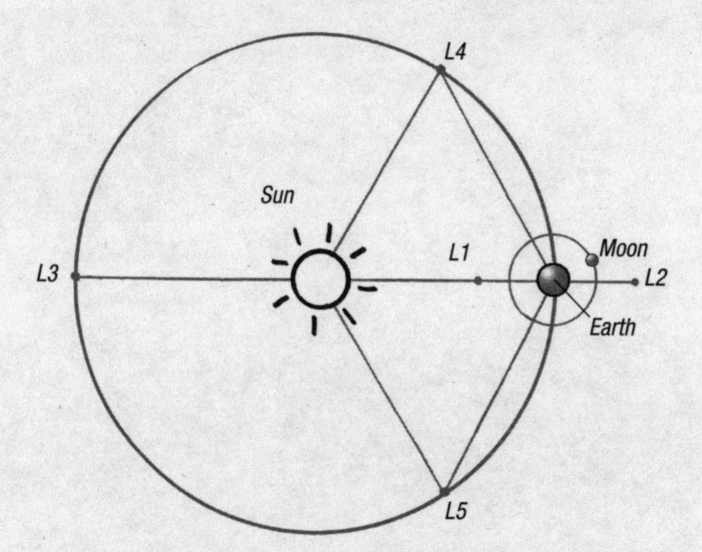

The Lagrange points of the Earth–Sun system, where the gravitational effects of the Sun and Earth cancel out to some extent.

Meanwhile, the greenhouse effect could actually be a boon on Mars, melting the planet's ice reserves to give running water. It has even been proposed that an asteroid laden with the greenhouse chemical ammonia could be steered into Mars. Others regard such ecological tinkering on other worlds with dismay. Then again, if we cannot fix the climate of our own planet sometime soon we may well have no choice.

CHAPTER 8

How to launch yourself into space

- Early ideas
- Rocket science
- How much fuel?
- Beating gravity
- Multi-staging
- Atmospheric re-entry
- Space tourism

Space is just an hour's drive away, 100 km (60 miles) above your head. And yet only a few hundred humans have ever been there. That's because, while possibly the most alluring destination for the intrepid traveller, it's also one of the most difficult places to reach – demanding the ride of a lifetime aboard a giant firework travelling at 25 times the speed of sound. If that doesn't put you off then check your bags and climb aboard the orbital express …

Early ideas

Going into space is one of humankind's oldest dreams. The idea began to edge its way closer to reality in 1903, when Russian space scientist Konstantin Tsiolkovsky published *The Exploration of Cosmic Space by Means of Reaction Devices.* In it, he described how human explorers could escape Earth's gravitational field to enter orbit around the planet and possibly even venture further afield. Tsiolkovsky imagined that we would use rockets to get there.

Rockets had already enjoyed a long, if less than peaceful, history within the confines of the planet's atmosphere. In the 9th century, Chinese scientists invented gunpowder and were quick to use it as a power source to hurl projectiles at their enemies. Ever since, rockets have been used in conflicts around the world, right up to the war in modern-day Iraq.

Rocket science

Rockets work by ejecting fuel exhaust at high speed, which accelerates the body of the rocket in the opposite direction. This is based on Newton's third law of motion, which says that for every force there is an equal force pushing in the opposite direction – when I fire a rifle the bullet is accelerated forwards out of the barrel, while the stock of the gun kicks back against my shoulder. Newton's third law means that as the rocket fuel burns, expands and is forced out of the rocket engine, an equal and opposite force is exerted on the rocket itself, which accelerates it forward.

The increase in speed is given by a principle known as the conservation of momentum. Momentum can be thought of as the impetus that a moving object has. Physicists calculate the momentum of a moving object simply by multiplying together its speed (measured in metres per second) and its mass (in kilograms). Fast, heavy things carry more momentum than slow, light things – which is why getting hit by a lorry hurts more than being bumped by a shopping trolley. The conservation of momentum says that the total momentum in any physical process must always stay the same. So if two billiard balls collide, the sum of both balls' momentum before the collision must equal the sum of their momentum after the collision. If one ball starts at rest, and is hit by a second ball which is stopped dead in the collision, then the first ball must carry away exactly the same momentum that the second came in with. And if they are both of equal mass then the second ball must leave with the exact same speed too.

How much fuel?

In rocketry, the conservation of momentum says that the rocket must gain momentum at the same rate as momentum is being carried away by the exhaust gas vented from the back. For example, say a small rocket spits out a kilogram of exhaust gas at 2,500 m/s. If the rocket body weighs 10 kg, that means the rocket will be moving forwards after the burn at 250 m/s. The momentum of the rocket must be the same as that of the exhaust, which is just the speed of the exhaust (2,500) times its mass (1).The speed of the rocket is this figure divided by its mass (10).

In fact, this is a slight simplification. It assumes the mass of the rocket is constantly 10 kg. But its mass is actually changing continuously as fuel get burned. Initially the rocket weighs 11 kg because it has to carry all the fuel it needs for the burn – and this extra baggage makes the final speed slightly less. Konstantin Tsiolkovsky worked out a mathematical equation for determining the speed that a rocket can reach given the weight of the rocket body, the weight of propellant and the speed of its exhaust gas, taking into account the fact that the propellant burns gradually. In our example above, Tsiolkovsky's rocket equation reveals that the rocket would actually accelerate to 238 m/s. Scientists call this total increase in velocity brought about by burning a given mass of fuel 'delta-*v*' – which comes from the mathematician's shorthand 'delta', meaning 'a change in', and the symbol for velocity, '*v*'.

Beating gravity

So how much delta-*v* does it take to get a rocket into space? The answer to that question was provided centuries ago by the English physicist Isaac Newton. In 1687, Newton published his theory of gravity. It was a monumental achievement for a scientist working in the 17th century: a single mathematical law that at a stroke explained the orbit of Jupiter and how apples fall from trees in Cambridge – phenomena separated by hundreds of millions of kilometres. Newton's theory was a universal law of gravitation, revealing not only why objects fall downwards in a gravitational field, but also how the planets circle the Sun. The mathematical description of orbits had been worked out 80 years earlier by the German astronomer Johannes Kepler. Newton's theory neatly provided the physics underpinning all three of Kepler's 'laws of planetary motion'.

However, Newton's theory also revealed how fast a rocket needed to travel in order to reach orbit around Earth. Launch a projectile into the air and it arcs through the sky before falling back to the ground. Launch the projectile faster and the arc carries it higher and further. Orbit is achieved when the projectile is travelling so fast that the curved surface of the planet falls away at exactly the same rate the rocket falls under the attraction of gravity, meaning that the rocket circles around the planet continuously. (Some readers may have heard of the term 'escape velocity' – the speed that a projectile, such as a cannon ball, must be given in a single kick at Earth's surface in order to completely escape the planet's gravity. However, escape velocity doesn't apply to rockets because they burn their fuel gradually – theoretically, a rocket could leave the atmosphere travelling at any speed as long as it had enough fuel to keep accelerating.)

Newton's theory predicted that the minimum speed needed for a rocket to achieve a circular orbit around Earth is around 7,800 m/s (25,500 ft/s) – quite quick. In fact, the speed of the orbit itself is just part of the equation. After adding in air resistance, the energy spent climbing out of Earth's gravitational field and other losses, the speed needed to get to orbit is actually more like 9,400 m/s (30,800 ft/s).

Multi-staging

Imparting this much speed to a rocket is no mean feat. But Konstantin Tsiolkovsky came up with a neat trick to make life easier for the engineers – multi-staging. His idea was for the rocket to shed weight as it flew by jettisoning sections of itself that had served their purpose, such as empty fuel tanks. For example, his equations showed that for a rocket of fixed body mass and fuel load, and carrying a payload that makes up 0.1 per cent of the total launch mass, splitting the rocket into three stages (each weighing 10 per cent as much as the stage below it) would leave the payload ultimately travelling twice as fast as it would if the rocket were just a single stage.

Tsiolkovsky was right on the money with his multi-staging idea, which proved to be crucial for the mighty *Saturn V* rockets that carried the first human beings to the Moon in 1969. Without staging, the *Saturn V* would only have been able to muster a delta-*v* of around 5,900 m/s (19,300 ft/s) – insufficient to get it to orbit, let alone the Moon. The *Saturn V* made use of so-called 'serial staging', where stages are burned and ejected one after the other. The other variant is 'parallel staging', where two or more of the stages are burnt simultaneously and then jettisoned. The Space Shuttle's solid rocket boosters are an example of parallel staging.

Atmospheric re-entry

It's not just getting into space that's difficult. Getting back down again is no cakewalk either. The main problem is the heating effect caused as the spacecraft re-enters Earth's atmosphere. A tragic demonstration of just how deadly this can be was the destruction of the US Space Shuttle *Columbia* on re-entering Earth's atmosphere on 1 February 2003. Damage sustained to the shuttle during launch allowed hot gases to melt the structure supporting its left wing, causing the spacecraft to break apart killing all seven astronauts on board. Heating during re-entry is due to compression of the air in front of the spacecraft. It is the same effect that makes a bicycle pump get hot as the air inside is compressed. When the spacecraft comes in from orbit at a speed of over 7,000 m/s (23,000 ft/s) it literally squashes a layer of air in front of it, heating it to 1,600°C (3,000°F) – hot enough to melt iron. The Apollo spacecraft returning from the Moon were travelling even faster, heating their exteriors to 2,800°C (5,000°F). Just as well these spacecraft were expendable. They consisted of a conical capsule, the blunt wide base of which hit the atmosphere to spread the force of re-entry (the deceleration could reach up to 7G, making the astronauts feel seven times as heavy as they do at Earth's surface). The base was coated with a heat shield to prevent the rest of the craft from melting. Apollo's heat shield was an ablator – a material that isn't totally impervious to heat but instead burns very slowly, charring until pieces break off, carrying heat away and exposing a fresh layer of shielding beneath. Parachutes then deliver the capsule to a soft landing. For Apollo an ablative heat shield was fine because the spacecraft were not re-usable. But the Space Shuttle was. So a new heatshield system was designed for it using heat-resistant foam tiles that cover its underside. Unlike Apollo's tough shield, which was concealed during launch, the shuttle's tiles are fragile and exposed – and this proved to be *Columbia*'s downfall.

Space tourism

Until very recently, travelling into space was the preserve of a select few professional astronauts. But space tourism is about to become a reality. British entrepreneur Richard Branson's Virgin Galactic company is offering to carry members of the public into space for a cool $200,000. Virgin Galactic's spacecraft is called *SpaceShipTwo*. The prototype, *SpaceShipOne*, won the Ansari X Prize in 2004 for the first private manned space launch.

Unlike Apollo and the Space Shuttle, *SpaceShipTwo* does not go all the way to orbit. Instead it flies on a so-called suborbital arc, crossing the boundary into space at an altitude of 100 km (62 miles) and peaking at 110 km (68 miles) above the planet's surface before dropping back to Earth. The passengers on board enjoy about six minutes of weightlessness at the top of the trajectory. As this is not an orbital flight the speeds involved are much lower. Branson's rocket delivers a delta-*v* of about 2,000 m/s (6,500 ft/s). It doesn't take off from the ground, but instead climbs into the sky slung beneath a jet aircraft. At an altitude of 16 km (10 miles), the rocket is released – rather like an air-launched missile – and then fires its engine to take it into space.

The low speeds mean there's negligible heating when the spacecraft re-enters the atmosphere. No heat shield is needed and the G-forces are far less traumatic. Like the shuttle, *SpaceShipTwo* has wings to enable it to glide down to a controlled landing on a runway. Branson has stated that the ticket price is expected to fall dramatically after the first few years of operation and may ultimately drop as low as the cost of a luxury holiday on Earth. If that happens then we may well get to spend a few minutes in outer space.

CHAPTER 9

How to survive a lightning strike

- Deadly discharge
- Electric current
- Electrical resistance
- What is lightning?
- Where to shelter
- Out in the open
- What are the odds?

It is said that lightning never strikes in the same place twice. Tell that to Pennsylvania man Don Frick who in 2007 proved the pundits wrong when he was struck by lightning 27 years to the day after first being hit by this awesome force from the heavens. Amazingly, he survived again – quite an achievement when you're tangling with up to a billion volts of electricity and temperatures nearly six times hotter than the surface of the Sun.

Deadly discharge

Lightning strikes Earth 50 times every single second. In the US alone, lightning strikes cause damage estimated at $4–5 billion and kill 90 people annually. Each strike produces electrical currents measuring tens of thousands of amps, and a peak power of a terawatt: 1,000 billion watts, or about twice the rate of electricity consumption of the entire United States. The temperature around each strike reaches 30,000°C (54,000°F), causing the air to expand at supersonic speed to generate the ominous thunderclap that warns of the oncoming storm.

Electric current

Lightning is the sudden discharge of electricity from a storm cloud down to the ground – or to another storm cloud of opposite electric charge. Charge is the fundamental property of electricity and is measured in coulombs, C, after the pioneering French physicist Charles-Augustin de Coulomb. Electric charges create electric fields, which enable the charges to interact with one another over distance. The charges can be either positive (+) or negative (–) and the electric fields they set up cause the charge carriers to either repel one another, as is the case for two 'like' charges (++ or ––), or to attract, which happens when the charges are opposite (+–).

The most common charge carrier is a subatomic particle called the electron. It is normally found in the atoms from which all materials are made, where large numbers of electrons orbit around each atom's nucleus and determine, among other things, the atom's chemical properties – how it reacts with other atoms. Each electron carries a tiny negative charge equal to -1.6×10^{-19} C. That is, −1.6 divided by a 1 with 19 zeroes after it. In some materials, however, electrons leak from the atoms and slosh around between them to form a 'sea' of electric charge carriers. Materials in which this happens are known as conductors, of which metals are a prime example. The surplus of electrons inside a conductor means that electric charge is free to move around inside it. And this can set up what's called an electric current. Current is a measure of the amount of electricity flowing through a conductor, and is measured in amps, after the French mathematician Andre-Marie Ampère. An amp is defined as the amount of electric charge (measured in coulombs) flowing per second past a given point in a conductor. Because the charge on the electron is so tiny a current of 1 amp corresponds to a flow of 6.2×10^{18} (6.2 billion billion) electrons per second.

Electrical resistance

Current doesn't flow of its own volition but moves to or from concentrations of electric charge. This happens because of the way charges attract or repel one another. So a negatively charged electron will tend to drift away from a concentration of negative electric charge (because like charges repel) and move towards an area of positive charge (because opposite charges attract). This is referred to as an 'electromotive force', or emf. Sometimes also known as a 'potential difference', it is measured in volts – after the Italian physicist Alessandro Volta. Batteries are a source of emf .

How much current flows through a particular conductor, say a piece of wire, when it is connected up to a battery is given by another property known as electrical resistance. This is the opposition that a current experiences as it tries to flow through the conductor, as the electrons jostle and squeeze between the lattice of atoms from which it is made – rather like commuters at a busy railway station. Resistance is particular to different materials and is measured in ohms, after the 19th-century German physicist Georg Simon Ohm. He also came up with a mathematical relationship, now known as Ohm's law, revealing that resistance is simply given by dividing the voltage applied to a conductor by the current that this voltage produces.

Resistance is also the reason light bulbs work. The resistance of the filament inside the bulb causes it to get hot – as the electrons all trying to squeeze through it rub against one another and against the atoms in the material. The rate at which heat and light are generated is measured in watts (after Scottish engineer James Watt) and is just given by the current times the resistance of the filament squared. Light-bulb filaments generally have a high resistance to maximize the amount of energy they give off.

What is lightning?

Lightning happens when the undersides of clouds acquire a large quantity of negative charge, building an emf of hundreds of millions, and in some cases even billions of volts between the underside of the cloud and the ground. The reason this happens is thought to be all down to ice crystals. These tend to gather positive charge, and are then carried to the top of the cloud by swirling currents within it. At the same time, heavier pieces of ice and water sink to the bottom of the cloud, carrying with them negative charges. This process of separation induces a massive negative electrical charge on the underside of a storm cloud. As this charge grows, it attracts positive charges, causing an equal but opposite charge to gather on the ground below. The high electrical resistance of the intervening air stops the charges from coming together and cancelling out – until, that is, the accumulated charge gets so great that it overcomes the resistance in one almighty discharge.

The air's resistance breaks down because of a phenomenon called ionization, where the huge electrical forces literally rip electrons from their parent atoms, gradually turning the air into a conductor. The process begins gradually. Tendrils of negatively ionized air called 'leaders' begin to snake their way down from the bottom of the thundercloud towards the ground. At the same time, on the ground, the storm bashes electrons from atoms to make positively charged ions, which also begin wending their way upwards from high points such as trees, telegraph poles – and people. When a leader from the cloud and one from the ground finally meet, current can flow and lightning strikes.

Where to shelter

The first thing you might know about a storm on the way is the distant rumble of thunder. Sometimes you can see flashes of lightning, too. Light travels much faster than sound in air (300,000,000 m/s compared with 343 m/s), and counting the number of seconds between seeing the flash and hearing the rumble is a good measure of the distance between you and the storm – about a kilometre for every three seconds. Normally you can start to hear thunder when the storm is about 16 km (10 miles) away – the bad news is this means you're already within range of the lightning. If you're able to, the safest course of action is to get indoors. Don't think you're safe under shelters and canopies. Get inside a building, where a lightning strike should be conducted safely down to the ground. But being inside doesn't make you totally safe. Lightning can still be transmitted into your home via the electricity supply. So do not use any electrical equipment during a storm – most indoor injuries from lightning are sustained by people talking on the phone. If you're away from buildings during a storm, a different strategy is needed. If your car is near then get inside and shut the door. The rubber in your tyres will do little to stop an electric current that's powerful enough to make it down through hundreds of metres of thin air. But the metal bodywork, so long as you're careful not to touch it while you're in there, should protect you – acting like a 'Faraday cage'. British physicist Michael Faraday showed in 1836 that the electric charges within an enclosure made of conducting material will always cancel out – and if there are no charges there can be no dangerous currents. It's the same mechanism that protects you if you're in a plane that gets struck by lightning. The average commercial jet aircraft gets struck about once per year – just as well its aluminium skin is designed to withstand currents of up to 200,000 amps.

Out in the open

You are at your most vulnerable when you're out in the open. Your main concern is to stay low – literally. Lightning takes the path of least resistance, travelling through the shortest distance of air possible, which is why it always strikes the highest point above ground. Sheltering from the rain under the only tree for miles around is a very bad idea (the same applies to standing near telegraph poles and metal fences). Not only do you risk severe burns should the tree be struck, but there is also the danger that you will get electrocuted directly thanks to a phenomenon called 'side flash'. This is where the current from the lightning bolt passes directly into the ground and spreads horizontally outwards. The electric charge this creates in the ground diminishes rapidly with distance. However, you may still receive a fatal surge of current through your body if you are standing close enough to where the lightning struck.

Standing in a dense forest of trees all more or less the same height is fine. But if your friends insist on sheltering under a lone tree, you could remind them that getting wet will actually boost their survival chances. This is because of a phenomenon called external flashover, where the current passes over the victim's body rather than through it – thus reducing the risk to the heart, brain and other organs. External flashover happens because water is an extremely good conductor, with a much lower electrical resistance than the human body.

The best advice in open country is to crouch down (keeping as low to the ground as possible), on the balls of your feet (minimizing your contact area with the ground), and with your feet close together (to lessen the risk from side flash). Some experts even recommend putting your hands over your ears and shutting your eyes (hearing and vision injuries are common among lightning strike victims). And should you feel the hairs stand up on your neck – a strong sign that a strike is imminent – hold your breath (some victims sustain internal burns from inhaling the superheated air). Oh, and make sure you and your friends all keep your distance while you're crouching – lightning can hop between people standing up to 6 m (20 ft) apart. Even once a storm seems to have passed, don't be lured into a false sense of security – charge remains in the air and lightning can still strike for up to half an hour afterwards.

What are the odds?

It seems incredible that a human being can stand in the path of such a powerhouse of natural energy and survive. And yet 90 per cent of lightning strike victims do. Injuries can be severe and debilitating – including burns, amputations and psychological trauma – but fatalities are not the norm. Statistically, men are more than four times more likely to be struck than women (maybe because men are more likely to be out in a storm in the first place). The average lifetime odds of being struck by lightning are about 1 in 3,000. For poor old Don Frick, it can be little consolation to know that by getting struck twice in the course of his life so far, he's managed to overcome odds of around 9 million to 1.

CHAPTER 10

How to cause a blackout

- Starfish Prime
- Induction
- Maxwell's equations
- How an EMP works
- E-bombs

There are few things that can match the devastation caused by the heat and explosive force of a nuclear weapon. But in the 1960s, the true power of what, up until then, had been considered a mere side-effect of a nuclear blast became apparent. Called an electromagnetic pulse, it has the power to destroy electrical equipment thousands of kilometres from the site of the explosion. With our reliance today on electronic communication, a well-placed electromagnetic pulse could bring an entire continent to a standstill.

Starfish Prime

On 9 July 1962, the American military carried out a nuclear test with a difference. They detonated a 1.4 megaton nuclear bomb in space, 400 km (250 miles) above the Pacific Ocean. The purpose of the test, called Starfish Prime, was to find out the effects of letting off nuclear weapons at high altitude. But with Starfish Prime they got more than they bargained for.

The bomb went off at just after midnight local time. Almost instantly, street lights started to go out in Hawaii nearly 1,500 km (900 miles) away. Burglar alarms were triggered and telephone networks were thrown into disarray. Hawaii had been hit by what's called an electromagnetic pulse (EMP). The effect had been known about since the first nuclear tests in 1945. Enrico Fermi – one of the scientists involved in the Manhattan Project to develop the American atomic bomb – had predicted that there would be an EMP from a nuclear explosion. As a result, all the electrical equipment involved in the early tests was shielded. But no one had anticipated the magnitude or the range of the effect as revealed by the Starfish Prime high-altitude test. With the impact of the EMP that hit Hawaii, scientists realized that this was not just a by-product of nuclear explosions – it was a powerful effect that could be weaponized with devastating effects.

Induction

EMP is caused by the intense electric and magnetic fields that are given out by nuclear explosions. The fields induce voltages in any electrically conducting materials that they pass through. If the field is strong enough, the voltage generated can be sufficiently high to overload circuits and wreck electrical equipment. When Enrico Fermi predicted this effect, he drew upon a theory that had been established 80 years earlier by Scottish physicist James Clerk Maxwell. Maxwell's theory drew together two phenomena in physics that had previously been believed to be distinct from one another: electricity and magnetism. Electricity is the force that makes electrical charges move to create an electric current. Electrical charges can be positive or negative and tend to flow towards areas where the charge is opposite – negatively charged electrons will flow away from the negative terminal of a battery and towards the positive terminal. Magnetism, on the other hand, is the force that makes compass needles move. Permanent magnets – of the sort that you might stick to your fridge – are made from types of metal with a property called ferromagnetism. This means that they are particularly susceptible to the effects of magnetic fields and, if left in one for long enough, become magnetized themselves. Permanent magnets have two oppositely magnetized poles, labelled north and south. Like electric charge, opposite poles attract and like poles repel.

By the early 1800s, it was becoming clear that these two concepts weren't as unrelated as physicists had believed. For example, it had been noticed that a current can be generated in a wire that's moving through a magnetic field. Conversely, when a current flows through a wire a magnetic field is produced around the wire. This phenomenon is known as induction. The laws governing how it works were formulated by British physicist Michael Faraday and the American Joseph Henry in 1831. Induction plays a central role in the operation of dynamos, which use an arrangement of magnets to turn rotational motion – for example, produced by a wind turbine – into electricity; and electric motors, which use an electrical current to turn a magnetized axle, so generating motion. The interplay between the electric field, the magnetic field and motion in each case were given by Fleming's rules, named after British scientist John Ambrose Fleming.

Maxwell's equations

It was starting to seem like electricity and magnetism were just different aspects of the same underlying phenomenon. Maxwell, along with his colleagues, provided the solid theoretical basis to back up this hunch by pulling all these emerging ideas about electric and magnetic fields together into a unified theory. In the 1860s, he carried out research that led to four equations, which summarized how electric charge, electric current, electric fields and magnetic fields are all interwoven with one another. The theory became known as electromagnetism, and the four equations will forever bear the name of the man who pioneered them: Maxwell's equations.

One of the most ground-breaking concepts to drop out of the equations is that light is an electromagnetic wave. It's made up of waves of both electricity and magnetism vibrating at right angles to one another. The resulting 'electromagnetic radiation' forms a spectrum depending on frequency with light sat roughly in the middle, at a frequency of around a million billion Hertz (where 1 Hertz, Hz, is one wave cycle per second). Below light in the spectrum are lower-frequency radio waves and infrared radiation, while above it are ultraviolet light, X-rays and, at the far end, high-frequency gamma rays. The frequency of an electromagnetic wave is directly linked to its energy and so radio waves have relatively low energy, while at the opposite end of the spectrum gamma rays are highly energetic – so much so they are hazardous to living things, and are classified as harmful radiation alongside alpha and beta particles (see *How to turn lead into gold*).

In 1897, Irish physicist Joseph Larmor used Maxwell's equations to prove another interesting fact. If you accelerate an electric charge, it gives off electromagnetic radiation. This is how a radio transmitter works. Passing a time-varying current through an antenna causes electrons in the antenna to vibrate in response to the time-varying signal. Let's say the current is the sound of someone's voice, which has been turned into an electrical signal using a microphone. The current makes the charged electrons vibrate and emit electromagnetic radiation – radio waves – and the time-varying signal in the current is imprinted into the radiation's waveform.

The waves can then be picked up by a radio receiver. Electrons in the receiver antenna are made to vibrate in synch with the waveform of the arriving radio waves. This sets up a time-varying current in the antenna (identical to the one in the transmitter antenna just seconds before) and this carries the signal to an amplifier, which cranks it up enough to drive a loud speaker through which the original voice message can then be heard.

How an EMP works

The EMP from a nuclear explosion is a kind of intense electromagnetic wave. Any conductor that it meets – such as the circuitry in an electrical device – acts like an antenna and absorbs some of the wave, which in turn sets up an electric current. If the current is strong enough it will overload the electrical circuit, burning out components and rendering it useless. Back in the 1940s, '50s and '60s the kind of circuits used in electrical devices were less sensitive and better able to withstand EMPs. But modern-day circuits, such as those inside computers, use tiny electrical currents inside semiconductor devices such as microchips. These circuits are easily fried by the massive currents induced by an EMP – which is why EMP weapons can have such a devastating effect on the digital infrastructure that civilization today relies on.

The EMP generated in a nuclear blast is divided into three components – called E1, E2 and E3. E1 is caused by gamma rays. These rays have so much energy that they knock negatively charged electron particles from atoms in the air that they collide with. These electrons streaming through the air set up a massive electric current. The current is accelerated by Earth's natural magnetic field, and these accelerated charges emit a pulse of electromagnetic radiation. E1 is the quickest and most destructive kind of EMP from a nuclear detonation. It generates high voltages that can wreck computer equipment, and cannot be blocked by standard surge protectors. E2 is caused by scattered gamma rays from the blast, which then collide with electrons created during the formation of the E1 pulse, accelerating them and releasing another EM wave. It's similar to the EMP generated by a lightning strike and so is easier to guard against. Finally, the E3 pulse is brought about by the disturbance to Earth's magnetic field caused by the explosion. This is similar to the electrical effects that are generated by solar storms. It can last for anything up to a few hundred seconds after the explosion and sets up currents in power lines that can damage transformers and electricity distribution networks. EMP effects can be reduced by shielding sensitive systems. However, a recent report for the US Congress into the vulnerability of the US to EMP attack stated that it was impossible to protect both military and civilian electronic systems against the most powerful types of EMP weapon.

E-bombs

It doesn't necessarily take a nuclear explosion to create an electromagnetic pulse. Following the invasion of Iraq in 2003, there was speculation that US missiles and war planes equipped with non-nuclear EMP weapons had been deployed. This came after Baghdad suffered power cuts during raids, even though generators and distribution systems appeared physically intact. Scientists have known how to make non-nuclear EMP devices for over half a century. They work using a conventional explosive charge to compress a magnetic field to high intensity. A current is passed through a coil of wire, which – through induction – then generates a magnetic field through its centre. Inside the coil is a hollow metal tube surrounded by a layer of high-explosive. The shock wave unleashed as the explosive detonates compresses the metal tube and the magnetic field inside it, generating an intense burst of electromagnetic radiation. Fields hundreds of times more powerful than the huge magnets used inside medical scanners (and tens of thousands of times greater than the field of a fridge magnet) have been generated using this method.

E-bombs like this could be mounted in the nose cone of a cruise missile or in a bomb dropped from an aircraft. It would deliver a localized but intense electromagnetic pulse, enough to disable communication systems, radar stations and vehicle electronics, and even penetrate inside shielded underground bunkers. Military pundits also speculate that EMP grenades may have been developed, to be used by infantry soldiers to knock out enemy electrical systems on the battlefield. In an age when almost every aspect of our lives relies on communications, computers and other delicate electronic systems, electromagnetic pulse weapons have the potential to destroy civilization without toppling a single building or claiming a single life.

CHAPTER 11

How to make an invisibility cloak

- Natural camouflage
- Stealth tech
- Invisibility cloak
- Hiding spacecraft
- Metamaterials
- Not being seen

Harry Potter's invisibility cloak gets him out of many a scrape, but could you ever build such a camouflage garment for real? A working invisibility cloak wouldn't just be a boon for young wizards, but for the military, in medicine and for anyone hoping to pull a sickie without getting spotted in town by their boss. The good news is physicists now think they can do it.

Natural camouflage

Camouflage is found throughout nature. Moths, tigers, even some fish that look virtually transparent underwater, are all making use of techniques that make them less visible, in turn helping them to evade predators or creep up on prey. Humans are no exception. We've become adept at hiding ourselves. Soldiers, for example, have perfected the art of blending in with their surroundings. But simple camouflage isn't quite what we mean by invisibility – the ability to completely disappear from view. Stage magicians have long made use of cunning arrangements of mirrors to create the illusion of invisibility. In practice, though, no one wants to be rattling around with glass plates bolted to their body. It would hardly be very stealthy.

Stealth tech

Invisibility of a sort has been in use by the military since as far back as World War II, when aircraft designers began building military planes that were invisible to radar. The German Horton Ho 2-29 was a 'flying wing' style aircraft coated with a radar-absorbing mixture of wood glue and charcoal. Luckily for the Allies, the war ended before the Germans could put the plane into service. In the 1960s, American engineers rolled out their first stealth aircraft, the Lockheed SR-71 Blackbird. This was a spy plane that used basic stealth technologies to surreptitiously photograph Soviet bases.

Today, the most famous stealth aircraft is the B-2 Spirit stealth bomber. Despite its wingspan of 50 m (160 ft), to a radar installation it looks no bigger than an aluminium marble. It achieves this first of all by its shape. The B-2's angles are designed so that a radar beam coming in from any direction will not be reflected directly back to the source – which would mean detection for the plane. Instead, it's rather like a disco glitter ball, scattering the radar beam off in all directions. Designers do this by carefully shaping the surfaces and by avoiding internal right angles – a radar beam striking a right-angled corner from any direction will be bounced right back to its source (just like a squash ball hit into the corner of the court will come right back at you). For this reason weapons and engines are all mounted internally. On some stealth fighter aircraft, even the pilot's head can cause an unwanted reflection, which is usually solved by coating the canopy glass with a thin layer of reflective gold. The exact shape that minimizes radar signature is hard to calculate, and so must be done using computers. The fact that the first mainstream stealth aircraft – the F117-A Nighthawk stealth fighter – had an angular polyhedral shape is largely because computers in the 1970s were only powerful enough to model the radar return from flat surfaces, with none of the sleek lines that would later grace the B-2. The weird shape of stealth planes makes them inherently unaerodynamic, and therefore quite unstable to fly. They get round this by using computers again, this time on board to constantly adjust the flight surfaces – flaps, rudder and so on – to stop the plane from careering out of control.

Shape isn't the only consideration. Like the early Ho 2-29, modern stealth aircraft are coated with radar-absorbent paint. Rather than wood glue and carbon, though, this contains tiny iron balls that soak up the radar energy, turning it into heat which can then be lost to the air rushing over the plane. Some stealth aircraft also add cool air to the engine exhaust to reduce their heat signature and even chemicals to minimize the formation of water vapour, which itself can have a strong radar signature. With all this technology on board, a single B-2 costs $2.67 billion. It is literally worth more than its weight in gold.

Invisibility cloak

Some early attempts to make aircraft stealthy involved putting lights on them to try and match their brightness to that of the background sky. Military pundits predict that as wars against technologically inferior opponents become more common, modern military aircraft could go the same way – as visual camouflage becomes more important than radar invisibility. This wouldn't be done using simple lights though, but flat-panel display screens to show an image on the aircraft's exterior of what's behind it, so rendering it virtually invisible to any onlookers. Similar technology has already been demonstrated by researchers at the University of Tokyo, who have made Harry Potter-style invisibility cloaks. Images of what's behind the wearer are relayed from video cameras to a projector, which displays them on the cloak's silvery fabric. The results are impressive, reducing the wearer to little more than a ghostly outline with objects behind them clearly visible.

The Tokyo team's set-up is crude. But the technology is set to improve drastically as cameras become ever smaller (think of the size of the camera in your mobile phone) and display screens become thinner and more flexible. For example, the Media Lab at MIT in Boston (and, in fact, many other groups around the world) are developing screens so thin and flexible they're known generically as 'electronic paper'. This is a kind of invisibility that exists in practice – not just in theory – in the world today. Of course, pedants might assert, however, that this is merely an illusion. Can we do better?

In H.G. Wells's novel *The Invisible Man*, the scientist Griffin uses a cocktail of chemicals to change the refractive index of his body to that of air. Refraction is a property of substances caused when their density slows down the passage of light through them. This makes the light tend to bend towards the 'normal' (a line perpendicular to the interface between two media – see opposite page) as it moves into a denser medium, and away from the normal as it moves into a medium that's less dense. In reality, Wells's scheme for invisibility is unworkable – at least for living things – as

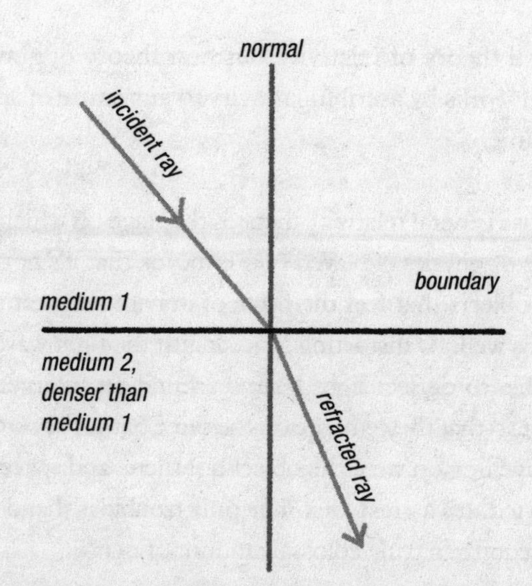

Refraction causes a light ray to bend towards the normal as it moves into a denser medium.

changing the refractive properties would mean altering the chemical properties of living tissue, which isn't likely to do the tissue's owner all that much good.

Hiding spacecraft

A more promising route is to try to deflect the light around the object that you want to make invisible. But how? In his book *The Physics of Star Trek*, US scientist Lawrence Krauss tries to explain the cloaking devices that are used by the Klingon and Romulan alien races in the show to render their interstellar battle cruisers invisible. Krauss's explanation makes use of Einstein's general theory of relativity, our best theory of gravity, which works by ascribing gravity to curvature of space and time.

Because general relativity dictates the stage on which the whole of physics is played out, it means that it's not just solid objects that feel the force of gravity but beams of light as well. By distorting space in just the right way, it is possible to deflect light beams around an intervening object so that they emerge on the same paths they would be travelling on were the object not there, and space not distorted. It's a great idea. The only trouble is that doing this requires a truly colossal amount of matter.

In 1919, British astronomer Sir Arthur Eddington carried out one of the key tests of general relativity, which involved measuring the degree to which starlight passing close to the Sun is bent by its gravity. Eddington confirmed Einstein's prediction that light grazing the Sun's surface is deflected by a minuscule 1.75 seconds of arc – less than one two-thousandth of a degree. And that's by the entire mass of the Sun. Even 'gravitational lensing' – the phenomenon exploited by astronomers to see quasars at the edge of the visible Universe (see *How to see the other side of the Universe*) requires the mass of an entire intervening galaxy cluster to do the bending. That's many hundreds of billions of Suns! Certainly not something the average Romulan spacecraft could fit in its cargo hold.

Metamaterials

Perhaps the most promising attempt at proper invisibility lies in the research of British physicist Professor John Pendry. In 2006, Pendry together with colleagues from the United States put forward a theoretical scheme whereby light could be channelled around an object by a special kind of material that acts in much the same way as the gravitational system envisaged by Lawrence Krauss – so that the light emerges on the other side travelling on the same path it would be on were the object and the cloak not there. Pendry likens it to water in a stream flowing around an obstruction.

His hypothetical substance is called a metamaterial – a material that has been carefully engineered on the smallest scales to have a specific set of properties that cannot be found in nature. Pendry's metamaterial would have to be very special, refracting the light away from the normal despite being optically denser than the surrounding air (remember, optically dense media typically refract the light the other way, towards the normal). This strange property is known as negative refractive index. In 2008, a multinational team of researchers actually succeeded in fabricating the first real examples of these weird, light-bending metamaterials. They are made by perforating pieces of silicon with a carefully designed pattern of holes – each just a hundred nanometres in diameter (about one ten-thousandth of a millimetre). These holes guide the light through the silicon in just the right way to give the desired negative refractive index.

So far, the team have only demonstrated two-dimensional metamaterials, and tiny ones at that – just a few billionths the size of a postage stamp. The next step will be a metamaterial that can create a negative refractive index in all three dimensions. Team member Dr Jensen Li believes this will be possible in the next few years.

Not being seen

A working invisibility cloak would have obvious military applications, to camouflage planes, tanks, ships and individual soldiers. And whereas existing stealth systems act specifically on waves of a particular frequency such as radar or light, metamaterials work at all frequencies simultaneously. There are peaceful applications aplenty as well. Don't like those plans to build a wind farm overlooking an area of natural beauty? Clad them in some of Professor Pendry's metamaterial and you'll never know they're there – until you accidentally blunder into one, that is. Pendry even believes there could be medical applications. A metamaterial cloak could be made to deflect not just beams of light, but any kind of electromagnetic wave – including strong magnetic fields. And so, for example, in MRI scanners – where it's not possible to have any metal objects or instruments present because of the magnetic fields involved – these objects could be wrapped in metamaterial to block their interaction with the magnetism. And, as with any breakthrough area of scientific research, there will be umpteen other applications too, many of which no one will have seen coming.

CHAPTER 12

How to be everywhere at once

- Young's experiment
- Why bands?
- Coherent light
- Monochrome vision
- Lasers
- Quantum double slits
- Schrödinger's equation

In the busy 21st-century world, many of us might wish we could be in more than one place at the same time. For subatomic particles this isn't a problem. According to the abstruse laws of quantum mechanics, particles of matter also behave like waves – enabling them to be not just in two places at once, but everywhere. One simple experiment gave physicists the insight they needed to unravel the laws of quantum theory.

Young's experiment

The double-slit experiment is probably the most startling demonstration of quantum physics. It was first carried out before quantum theory was even a twinkle in the eye of Max Planck, Einstein and colleagues. Nobel prize-winning US quantum physicist Richard Feynman would later remark that pretty much every aspect of quantum physics is encapsulated by this astounding experiment. So what is it?

British physicist Thomas Young was the first person to perform this experiment in 1801. Young was trying to figure out whether light is made of particles or waves. To do this he shone a beam of light onto a screen with a pair of narrow slits cut into it. The light passed through the slits to illuminate a second screen, this time with no slits. Young postulated that if light was made of particles then the second screen would be evenly illuminated by the light from the two slits. But if it was made of waves there should be a pattern of bright and dark bands, known as 'interference fringes'. And this second possibility is exactly what Young observed.

The fringes form as the light waves from each slit collide with one another at the second screen and overlap, rather like the way ripples on water can overlap. When this happens on water, the two wave-forms add together. So where the peaks of two waves coincide, a large peak is formed (called constructive interference); where the dips, or troughs, of two waves meet, a deep trough results (constructive interference again); and where a peak and a trough of equal size meet, the two simply cancel one another out (known as destructive interference). This principle, in which waveforms add together, is known as superposition.

Exactly the same thing happens with light rays at the screen in Young's experiment. A ray of light is a wave, like a wave on a string, with peaks and troughs. Where a wave peak in the light ray from one slit coincides at the screen with a peak in the light ray from the other, a bright fringe is formed. And the same thing happens where two troughs meet. But where a peak and a trough coincide the two light waves cancel one another out to form a dark band.

Why bands?

Bands are formed because the light from one slit falling on the second screen will generally have travelled a different distance from the light from the other slit. Right in the middle of the second screen both light rays have travelled exactly the same distance, so their peaks and troughs both overlap exactly. There is thus constructive interference, resulting in an overlapping bright image of both slits – i.e. a bright band. Look to either side of the bright central band and you reach a point where the ray from one slit has travelled half a wavelength less than the other. These rays then interfere destructively, resulting in overlapping dark images of each slit – i.e. a dark band.

Coherent light

Of course, this all assumes that the light waves were all in lockstep with one another when they passed through both the slits. In other words, a wave peak passes through one slit at the same time as a wave peak passes through the other, while a trough passes through one at the same time as a trough passes through the other. Physicists call light that has this property 'coherent'. If, on the other hand, the light was incoherent, its waves would be hopelessly out of synch and there would be no chance of witnessing any interference fringes. Most light sources in nature are incoherent. When Thomas Young first performed the double-slit

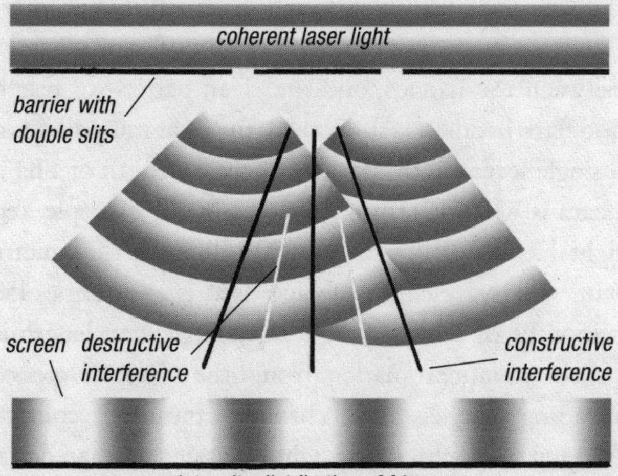

Shown here is the double-slit experiment. When coherent light is shone through two slits and onto a screen, a pattern of bright and dark fringes is formed as peaks and troughs in the waves travelling from each slit meet and interfere with one another.

experiment in the early 19th century, he was able to generate reasonably coherent light by shining an incoherent source through a pinhole. Because the pinhole is to all intents and purposes a dot of zero size, it filters out the incoherent variations from point to point along the light's wavefront. But this wasn't the only demand placed on the light used in Young's experiment. It was also crucial that the light should be made up of waves of just one wavelength (see *How to make the loudest sound on Earth*).

Monochrome vision

If light of all different wavelengths is passing through the slits, each wavelength will make a different interference pattern on the screen, with different spacings between the fringes, and the clean pattern of bright and dark bands is lost. Instead, the light must all be of a single wavelength. Because the wavelength of a light beam is what determines its colour (for example, red light has a wavelength of 650 billionths of a metre while the wavelength of blue light is shorter, at 450 billionths of a metre), light of a single wavelength is called monochromatic (from the Greek 'mono', meaning single, and 'chroma', meaning colour). Making monochromatic light is easier said than done. Ordinary light sources – such as filament light bulbs – give off light at a large range of wavelengths. The range is determined by the temperature of the source, in this case the hot filament inside the bulb. German physicist Max Planck used quantum theory (also known as quantum mechanics) to work out the spectrum of radiation from a hot body. The spectrum is just a graph with the range of wavelengths of the electromagnetic radiation given off along the bottom and the intensity of the radiation at each wavelength plotted vertically. For a filament bulb most of this radiation is at infrared and visible light wavelengths, but there's a large spread either side.

Young overcame this problem by using light from a mercury vapour lamp. This is a source of monochromatic light that operates on the principles of quantum theory – although Young didn't realize it at the time, because the theory was yet to be formulated. It works using a glass bulb filled with mercury vapour, through which a large electric current is passed. Energy from the current is absorbed by the mercury atoms, which makes electrons in the outer shell of each atom jump up to a higher energy level. Energy levels are one of the key features of quantum theory, and their existence was one of the predictions of the Schrödinger equation which lies at the theory's heart (see p.116). At very small scales, the energy of an electron in an atom is only allowed to take one of a discrete range of values. An atom raised to a higher energy level soon drops back down again, re-releasing the energy as a packet of light – called a photon. The energy of the photon is determined by its wavelength, so electrons all dropping down from the same energy level will have the same wavelength – in other words, they are monochromatic.

Lasers

Nowadays, scientists perform demonstrations of the double-slit experiment using lasers. The word laser is an acronym, standing for 'light amplification by the stimulated emission of radiation'. A laser works in a similar way to a mercury lamp. A material known as the lasing medium – ruby is a common choice – is first pumped with energy from a source such as an ordinary flash tube. This raises electrons in the lasing medium to a particular energy level. As they start to drop back down, monochromatic light is released with a characteristic wavelength given by the gap between the energy levels involved. When a photon released by an atom in the lasing medium passes near to another atom with an electron in an energized state, it can trigger the electron to drop down and release another photon. Not only does this new photon have exactly the same energy and wavelength but the peaks and troughs of its waves are naturally synchronized, making it coherent too. This process is called 'stimulated emission', the theory of which was developed by Albert Einstein in 1917.

The laser was invented by US scientist Charles Townes in the late 1950s. It consists of a cylinder of the lasing medium with mirrors at each end to bounce photons back and forth inside it, so their numbers become amplified by stimulated emission. One of the mirrors is only half-silvered, allowing a proportion of the light to escape as a tightly collimated beam. Lasers proved to be one of the greatest inventions of the 20th century, underpinning devices such as CD players, fibre optics and high-capacity data storage. Today, you can buy cheap lasers as pointers and even key fobs. You can even recreate Young's experiment by plucking a hair from your head and shining a laser at it. The hair acts as the obscuring gap between the two slits, and on the wall behind it you will see a pattern of bright and dark fringes.

Quantum double slits

When Young produced interference fringes in 1801, it seemed like fairly conclusive evidence that light is a wave. But here's where it starts to get really interesting. In the late 19th century and early 20th century, experimental observations began to roll in suggesting that light can still exhibit some degree of particle-like properties. Max Planck's accurate explanation of the radiation from hot bodies relied on this – as did Einstein's explanation of the photoelectric effect (see *How to harness starlight*). So what was going on? Surely light had to be either one thing or the other?

To get the bottom of the mystery, experimentalists resolved to repeat Young's experiment – but with a twist. This time, rather than shining a whole beam of light through the apparatus, they would send just one particle – that is, one photon – at a time. Common sense would dictate that this single particle of light can only pass through one slit or the other, and so it should be impossible to generate any kind of interference from light passing through both slits. What actually happens is quite remarkable. The scientists fired a photon through the apparatus and recorded the position of the resulting dot on the second screen. Then they repeated the process over and over. As time passed, and more dots accumulated on the screen, a pattern began to take shape – the original interference pattern. But how can this be? At any one time there was only one photon in the system. There's nothing else there for it to interfere with. The only way a single photon can give rise to interference is if the photon somehow interferes with itself. Put simply, it must pass through both slits – it is in two places at once.

Schrödinger's equation

This single experiment laid bare the inherent weirdness of the quantum world. Down at this level there is no such thing as a pure particle that is in one place at one time, or a pure wave that is spread out in space – just a strange mixture of the two. Today, physicists interpret the wave aspect as a wave of probability. Peaks of the wave correspond to where you're most likely to find a particle when you actually make a measurement. A probability wave passing a point in space makes it more likely that you'll find a particle at that point – in much the same way that a crime wave in your neighbourhood makes it more likely that a crime will be committed there. In the double-slit experiment, the probability wave passes through both slits and interferes with itself on the screen to create an interference pattern, the peaks of which are where each photon particle is most likely to be found. In 1926, Austrian physicist Erwin Schrödinger drew upon observations such as the double-slit experiment to construct a now-famous equation governing how a particle's probability wave behaves. And from this equation would follow the rest of quantum theory, one of the landmark achievements in 20th-century physics.

CHAPTER 13

How to live forever

- The universe next door
- Many worlds
- Decoherence
- Quantum suicide
- Quantum immortality
- Live long and prosper

From tales of the mythical fountain of youth to drugs that counteract the natural ageing process, the quest for immortality has long fascinated scientists and philosophers. It's a field of research that's usually more at home in the labs of medical researchers than the minds of theoretical physicists. But now a physicist has found a way by which it might be possible to live forever – by hopping between parallel universes.

The universe next door

Parallel universes make for great science fiction stories. But in real-world science, they were something of a nebulous concept for many years. During the 1920s and '30s – the years following Albert Einstein's formulation of the general theory of relativity – physicists discovered wormholes: bridges linking the space and time of our own Universe with those of others. But no one had much clue where these other universes actually were. Are they part of the same space and time as our Universe, or totally disconnected from it? Neither did anyone know whether they obeyed the same laws of physics as our Universe, or different ones. Would they contain galaxies, stars and planets like ours – and maybe even people – or are they barren and empty? Why are these parallel worlds even there in the first place? And that's if they exist at all – no one was even sure whether they were real or just a curiosity thrown up by the complex mathematics of Einstein's theory.

That began to change in 1957 when US physicist Hugh Everett came up with a new way of thinking about quantum theory, the physics governing subatomic particles. Quantum theory supposed that solid particles don't always behave like solid particles, but can also exhibit properties more reminiscent of wave motion. Particles, it seemed, could do things that it had been thought up to that point only waves can do – such as diffracting (spreading out) as they passed through narrow slits, and interfering with one another like ripples overlapping on the surface of a pond. Quantum ripples are interpreted as probability waves. The form of the wave describing a particle is known as its wavefunction: an undulating surface covering the whole of space, with the height of the undulations at any point giving the likelihood of finding the particle there when a measurement is made.

Up until the time of Everett's work, most physicists believed in the so-called Copenhagen interpretation of quantum theory, according to which the act of measuring a particle causes its wavefunction to 'collapse' – so that rather than a wave spread out through space the measurement reveals a solid particle with a definite location. In this view, quantum probabilities are simply due to the measurer's ignorance of the state of the quantum particle.

Many worlds

Everett threw out the notion of wavefunction collapse in favour of a new view that he called the relative state formulation, which became known as the many worlds interpretation. Rather than the wavefunction of a quantum system collapsing into just one of the many possibilities open to it, Everett suggested that there exist many universes where every possibility is played out for real. The particle's wavefunction doesn't just exist in our universe, but is spread out across this 'multiverse' of many worlds. The question then becomes not how the wavefunction collapses in our one Universe, but which one of the many universes is the one we each subjectively happen to find ourselves in. Here's an example. Let us say a particle has a very simple wavefunction that spreads the particle's position between one of just two points in space. The set-up is such that the particle has a 30 per cent chance of being in the first position and a 70 per cent chance of being in the other. In the Copenhagen view, measuring the particle would force its wavefunction to collapse onto one location or the other, with relative odds 3:7. In the many worlds view, however, the particle really does exist at both locations – in 30 per cent of all universes the particle is in the first position, while in the other 70 per cent it's in the second position. Making a measurement picks which universe we're in, and we see the particle's position as it is in that universe.

Decoherence

In the Copenhagen interpretation, collapse of the wave function marks the transition from thinking of the particle's behaviour as a wave (quantum behaviour) to viewing it as a solid particle (classical behaviour). An analogous process called decoherence happens in the many worlds interpretation. Whereas collapse of the wavefunction attributes this transition to the meddling fingers of the observer, decoherence puts it down to the inevitable interaction of a delicate quantum system with its classical environment. Prior to decoherence, the particle is in a purely quantum state. Quantum interference (see *How to be everywhere at once*) between the different universes of the multiverse means that we view its behaviour as a wave. But as soon as the delicate balance needed to maintain this quantum state is upset by outside forces, the system decoheres into a classical state. The universes of the multiverse then peel apart and we end up in just one of them.

Quantum suicide

The sprawling network of parallel universes that the many worlds interpretation relies on has made many scientists sceptical about the theory. In recent years, however, the meteoric rise of quantum computers (machines that harness the power of copies of themselves in parallel universes to carry out lightning-fast calculations – see *How to crack unbreakable codes*) have convinced many. But will it ever be possible to gather incontrovertible evidence as to whether parallel universes do or don't exist, and so prove which is right: Copenhagen or many worlds?

One scientist thinks such a proof is possible. Max Tegmark, a US physicist at the Massachusetts Institute of Technology, has come up with a macabre experiment that could tell between the two. He calls it 'quantum suicide'. It's an outlandish twist on the old Schrödinger's cat thought experiment, where a cat is locked in a box with a phial of poison and a radioactive source. If the source emits a particle of radiation, the poison is cracked open and the cat dies; otherwise it lives. Radioactive decay is a quantum process, meaning that the wavefunction of the radioactive source must be a mixture of decayed and not decayed. So – the reasoning goes – the cat must therefore be a mixture of both dead and alive at the same time before it's measured. Tegmark replaced the phial of poison in the experiment with a special kind of gun that fires automatically once every second. The gun is linked to a radioactive source. The default setting is for the gun to just click on an empty chamber each time – unless, that is, the source has emitted a particle of radiation in the last second, in which case the gun fires a live bullet. The source is chosen so that the probability of there being a decay event in any one-second interval is 50 per cent. To anyone watching, the gun makes a random sequence of clicks and bangs, each interspersed with equal frequency. But according to Tegmark, and if the many worlds interpretation is correct, anyone brave enough to put their head in front of the barrel will see the gun click every time with 100 per cent probability. It never fires a live round.

Tegmark's explanation is that in the many worlds view there are always universes in which the gun doesn't fire a live bullet. The wavefunction of the radioactive source is spread evenly across the multiverse – and, therefore, so is the final state of the human subject, including their very consciousness. Given that this consciousness won't exist in universes where the gun fires a bullet, it must therefore always find itself in one of the 50 per cent of universes where the gun just clicks. The experimenter must always perceive herself to end up in one of the universes in which she survives. From the point of view of an onlooker, however, the process wouldn't be too pretty. Copies of the experimenter and her consciousness exist in all the other universes in the multiverse. And in 50 per cent of these universes the gun will fire live bullets. The onlooker will see it click and fire at random, so sooner or later they will see the experimenter shoot herself – quantum suicide. For this reason, the only way this method could really prove many worlds to a sceptic would be for the sceptic to put their head in front of the barrel – which, given their scepticism, is unlikely to happen.

Quantum immortality

The astute reader may be able to spot where Tegmark is going with all this. Anyone killed instantaneously (not rendered comatose or just critically injured) as a result of a quantum event can conceivably take advantage of the fact that (in the many worlds view) there exist parallel universes where the quantum event has the opposite outcome and they therefore survive. They will always find themselves in one of these universes, and could therefore achieve an immortality of sorts.

The quantum nature of the fatal event is crucial. For example, if you cross the road while fiddling with your iPhone and get hit by a bus load of commuters from the local park and ride there's nothing quantum about that, and you will be ground into the asphalt with 100 per cent probability. Only when your death is triggered by the randomness of quantum physics will there be other universes where the randomness swings in your favour so that you can live. Tegmark points out that one such quantum killer is cancer, which begins with the mutation of a DNA molecule (molecules are made of atoms bonded together by quantum forces). He imagines a swarm of tiny nanorobots that could swim through your blood stream monitoring your DNA. As soon as they detected a cancerous mutation, the robots would kill you instantly – perhaps by remotely triggering a switch in your brain that instantly stopped all of your life processes. If he's right then by using this system you would never get cancer – quantum physics would provide a bizarre cure to this often fatal illness.

Live long and prosper

If you're going to enjoy your newfound longevity you'll need enough money put away for a lengthy retirement – to make your life not only long, but prosperous as well. And Tegmark has this taken care of too. Some lotteries now use quantum random number generators to pick the winning ticket. These use the randomness of quantum processes – such as radioactive decay – to select their numbers. This makes them more secure than other methods, which sometimes use computer algorithms to generate a seemingly random sequence that can be predicted by someone who knows which algorithm is being used.

Quantum randomness is completely unexploitable. Until, that is, you remember the many worlds interpretation. If the lottery numbers are determined by a quantum event, yours will always be the winning ticket somewhere out there across the multiverse. Linking the same random-number generator to a quantum suicide gun rigged to kill you if you lose ensures that's exactly where you end up. Don't try this at home!

CHAPTER 14

How to teleport

- Here nor there
- Measuring atoms
- Entanglement
- The quantum teleporter
- Beaming qubits
- Download yourself

It's the ultimate in personal transportation – a machine that can disassemble your body atom by atom and then beam all the information about you at the speed of light to a receiver, where you are promptly reassembled. If that sounds like the ramblings of a science nerd who's seen too much *Star Trek* then think again – teleportation has been verified experimentally.

Here nor there

The word 'teleportation' was coined in 1931 by the independent British researcher Charles Fort. He spent many years of his life gathering and cataloguing tales of the bizarre and the unexplained. While attempting to explain the reports of showers of stones, ice and occasionally even live animals from the sky, Fort supposed that 'teleportation exists, as a means of distribution of things and materials.'

But Fort also turned up mysterious accounts of human teleportation, such as that of Gil Pérez. The story goes that on 24 October 1593, the chief of the Guard at the Governor's Palace in the Plaza Mayor, Mexico City, noticed that one of his soldiers was not wearing the correct uniform. The dazed soldier said his name was Gil Pérez and that he was a guard at the Governor's Palace in the Philippine capital, Manila. While realizing he was no longer in the Philippines, Pérez had no idea how he had come to be in Mexico City. Probably the most famous teleportation legend is the story surrounding what has become known as the Philadelphia experiment. In 1943, during a supposed test of new 'invisibility equipment', the US Navy destroyer USS *Eldridge* is said to have disappeared from the Philadelphia Naval Shipyard to be spotted just moments later in waters off Norfolk, Virginia, by the crew of a passing merchant ship. The *Eldridge* allegedly teleported back to Philadelphia shortly after, where its crew suffered terrible after effects, including spontaneous human combustion and the melding of body parts with the ship. The US Office of Naval Research dismisses the experiment as an urban myth, probably arising from research on degaussing – the practice of passing currents around the hull of a ship to render it impervious to magnetic mines. It may be no coincidence that the writers Robert Heinlein and Isaac Asimov – two of the most creative minds in science fiction – both worked at the Philadelphia Naval Yard between 1943 and 1945.

Measuring atoms

For many years, it was believed that the laws of quantum physics made teleportation impossible. In 1927, the German physicist Werner Heisenberg had put forward his so-called uncertainty principle, which turned out to be one of the cornerstones of physics. In a nutshell, it said that it is impossible to know everything about a quantum particle – precision in your knowledge of one of its properties must be traded off against imprecision in one of the others. One way to think of it is that the act of measuring one property of the particle disturbs it so much that you lose all accuracy in the other.

But if this was right, it meant that measuring the state of every atom and molecule inside an object in order to teleport it was going to be impossible. All you'd be allowed is a partial description, and what you got out the other end of the teleporter might bear little resemblance to what went in. The writers of the *Star Trek* TV show even went so far as to invent a fictional device that they called the 'Heisenberg compensator' to give them a retort to the flood of letters they received from science-savvy fans.

Entanglement

In a groundbreaking piece of research published in 1993, a team of scientists in the United States realized that you don't have to measure all the information about a quantum particle in order to transmit it. In effect, quantum theory lets you teleport information that you don't actually know. The idea works using a phenomenon known as quantum entanglement. This is where you take two subatomic particles and bring them together in just the right way so that their properties – such as speed, momentum, energy and so on – become linked. Two entangled particles taken to opposite sides of the Universe can appear to exhibit a kind of faster-than-light communication between each other, since measuring the state of one particle instantaneously fixes the state of the other all those billions of light years away. Albert Einstein was aware of quantum entanglement and used it to express his general disdain for quantum theory, famously calling it 'spooky action at a distance'. William Wootters, one of the US team, likens the behaviour of entangled particles to two boxes – one has a blue ball in it and the other has a black ball in it, but you don't know which ball is in which box. Take the two boxes to opposite sides of the Universe, and open one of them – and what you see instantly tells you which ball is in the other box too.

Of course, quantum particles are specified by more than just one number, so as well as a blue ball and a black ball you might have some other pairs like a green ball and a red ball, a brown ball and a pink ball, and a white ball and a yellow ball. If you open one box and you find it to contain blue-red-pink-white then you know the other box is holding black-green-brown-yellow. You know this without having to make any measurements of (i.e. look inside) the second box. And this same sort of 'oppositeness' in their properties is a fundamental relationship between pairs of entangled quantum particles. One property particles have, for example, is called quantum spin. It's very loosely analogous to rotational spin of everyday experience. However, in the quantum world spin can take just one of two values – called 'up' and 'down'. The two particles in an entangled pair will always have opposite quantum spin. But until a measurement is made it's impossible to say which has spin up, and which has spin down. What makes entanglement really weird – and slightly different from the boxes analogy – is that nature itself doesn't decide which is which until one of the particles is actually measured.

The quantum teleporter

This is how it's possible to get round Heisenberg's uncertainty principle for teleportation. The uncertainty principle stops you from measuring the information stored on a particle, but entanglement means that you can still send that information even though you don't know it. It's rather like someone asking you to post a letter for them – you don't need to know the contents of the envelope in order to put it in the post on their behalf.

Here's how it works. Say you want to teleport a particle, A, from the surface of Mars up to the Starship *Enterprise*. You first need two other particles, call them B and C, which have been entangled together at an earlier time. B is with you on the surface of Mars, while C is already up on the *Enterprise*. Next, you join A with B and then make an ordinary measurement of the relationship between A and B. Although uncertainty prevents us from knowing the exact state of particles A and B, knowing the relationship between them is permitted and turns out to be all you need in order to teleport the state of A. The result of the measurement is then radioed ahead to the *Enterprise* where it can be used to extract an exact copy of A from particle C. For example, if the measurement showed that A and B are in opposite states to one another then since B and C are also opposite – because they are entangled particles – that reveals that A and C are already in exactly the same state. In other words the state of A has been transferred to C. What's more, the original particle A is destroyed in the process. A has effectively been teleported.

Beaming qubits

The discovery wasn't confined to humble pen and paper for long. In 1997, scientists at the University of Innsbruck were able to put the scheme into practice. They used an entangled pair of particles to teleport photons – particles of light – 2 m (6 ft) across their lab. Although the experiments aren't quite up to trips into space just yet, they demonstrate the reality of beating Heisenberg's uncertainty principle in practice. In 2002, a collaboration between the universities of Oxford and Calcutta put forward a modified version of the technique that can teleport not just light but solid particles of matter, including atoms and molecules. The first atoms were teleported experimentally by a US/Austrian team in 2004.

So if we can teleport atoms, how soon might it be possible to send larger objects and, maybe, people? Don't hold your breath. The teleportation experiments conducted so far have been largely about communicating information, such as the 'qubits' of quantum data that are processed on a quantum computer (see *How to crack unbreakable codes*). Qubits are fragile entities, suffering from a problem known as 'decoherence' – where interactions with their local environment can disrupt the delicate quantum balance by which the information they hold is encoded. And this means that they cannot be sent over conventional channels, such as by radio or electronically down a wire.

Download yourself

The real trouble with teleporting large objects is the sheer volume of information to be sent. An average-sized human is composed of roughly 10^{28} atoms – that's 10 billion billion billion times the largest units of matter that have been successfully teleported to date. If you assume that each atom can be specified by just one bit of information, a simple binary 1 or 0 (and that's a generous assumption – it takes more data than that to specify the state of an atom) then that's 10^{28} bits that need to be communicated from the transmitter to the receiver. The fastest internet backbones in service today can ship data at a fairly blistering rate – some 100 billion bits per second. But even at this not-inconsiderable rate, getting the information needed to specify the state of every one of Captain Kirk's atoms is going to take about 3 billion years – two-thirds the age of Earth.

The other difficulty is that teleportation machines would be needed at both ends of the process. So whereas the *Enterprise* can beam crew members in or out, from and to, pretty much anywhere, in reality only transportation between established destinations and jumping off points would be possible. Other researchers wonder what the implications for the self would be if you were to teleport yourself using quantum entanglement. After all, what comes out the other end isn't actually you – it's a copy built from a completely new set of atoms, while the original you is destroyed in the transmission process. Would you notice the difference? Would your brain structure be accurately preserved – would the new you have the same memories and psychological outlook as the original? And what would really happen if a fly got into the teleportation pod with you? Teleportation of even everyday objects – let alone people – is still many decades, probably even centuries, away.

CHAPTER 15

How to fit a power station in your pocket

- Ultimate energy source
- Dirac's discovery
- The asymmetric Universe
- Antimatter power
- Storing antimatter

Many a sci-fi yarn is spun around antimatter as the super-fuel of the future. Antimatter generates vast quantities of energy from the tiniest volumes of fuel. Now some NASA engineers are taking note. They've already produced designs for the antimatter-powered spacecraft that they hope will fly in the coming centuries. Others have suggested that antimatter mined from space could be a possible future energy source on Earth.

Ultimate energy source

Antimatter is what you get when you take a particle of matter and reverse key properties such as its electric charge. It's potent stuff. When a particle of matter meets its antiparticle the two annihilate, releasing their combined mass and energy in a flash of radiation. Converting mass into radiation in this way was one of the predictions of Einstein's special theory of relativity. The equation $E=mc^2$ links mass and energy by a fundamental constant of physics – the speed of light. The speed of light is very large – 300 million m/s – and the speed of light squared is even larger. This means that a very small amount of mass translates into a huge amount of energy. Indeed, annihilating just a quarter of a gram of antimatter would match the total energy output of the atom bomb that was dropped on Hiroshima in 1945.

Dirac's discovery

The discovery of antimatter is also thanks to Einstein – albeit indirectly. In the early 1900s, physicists were beginning to realize that matter on the smallest scales doesn't behave at all like matter in the everyday world. Bounce a ball on the ground and it accelerates and rebounds according to Isaac Newton's well-established laws of motion. But subatomic particles, it seemed, obeyed a new set of laws entirely.

In particular, it seemed like solid matter on these tiny scales has wave-like properties – more like a beam of light. Subatomic particles can be diffracted (spread out) as they pass through small apertures, and can interfere with one another just like waves. In the first few decades of the 20th century physicists began piecing together a mathematical description of this behaviour, which came to be known as quantum theory. One of the most significant developments came in 1926, when an Austrian physicist called Erwin Schrödinger published an equation describing the wave-motion of solid particles. Schrödinger specified the position of a particle in space by a 'probability wave' – with peaks of the wave corresponding to locations where the particle was most likely to be found. The Schrödinger equation, as it became known, revealed how this wave evolved in time. Enter British physicist Paul Dirac. As the Schrödinger equation stood it only described the motion of slow-moving particles. Dirac wondered whether it might be possible to formulate a version that was consistent with Einstein's special theory of relativity, taking into account the behaviour of objects moving at close to light speed. After a little mathematical gymnastics, he produced the Dirac equation – a relativistic version of Schrödinger's wave equation.

When Dirac applied his equation to electrons, he was surprised to find that it had not one but two solutions: one for the electron and one describing a particle with opposite electric charge. Dirac had predicted the existence of the positron – the antimatter partner to the electron. In 1932, just five years later, the first positron was detected experimentally. US physicist Carl

Anderson observed high-energy particles from space known as cosmic rays using a cloud chamber – an enclosure in which the air is saturated with alcohol vapour. When a high-energy particle passes through the vapour, it causes a trail of droplets to condense behind it. Placing a magnet over the chamber caused the trajectories of electrically charged particles to bend into a circle, and the radius of the circle revealed the particle's mass. Anderson observed plenty of electrons in his experimental set-up, but now and again he saw a particle with the same mass as the electron but curling in the opposite direction – signifying that it had opposite charge. Anderson had identified the first particle of antimatter.

The asymmetric Universe

The prediction and subsequent detection of antimatter was a resounding triumph for the emerging field of quantum physics, and for theoretical physics as a whole. But it brought with it a very big problem that physicists have still not been able to fully resolve – namely, why is our Universe made predominantly of matter and not antimatter? The standard model of particle physics says that the Big Bang, in which our Universe was born, should have generated equal amounts of matter and antimatter, which would have annihilated each other completely, leading to a Universe filled with radiation and nothing else – no planets, no stars and certainly no physicists. But our Universe is filled with matter. Observations have shown that the Universe contains around a billion photons (particles of light) for every baryon (large particles, such as protons and neutrons). Each annihilation of a baryon with an antibaryon in the early Universe will have produced one photon. So the implication is that there was a tiny imbalance of matter to antimatter – roughly a billion antimatter particles to every billion-and-one matter particles. In 1967, the Russian physicist Andrei Sakharov showed that the only way around this problem – known as the 'baryon asymmetry' – is through a phenomenon called CP violation. CP is a kind of symmetry in physics – short for charge-parity invariance. Symmetries are transformations of the laws of physics that leave the results unchanged. Initially particle physics was understood to be CP-invariant, which means that if you simultaneously reverse a particle's electric charge (that is switching matter for antimatter) and parity (its sense of left and right) then its behaviour remains unchanged. It's essentially a statement of Anderson's observation that positrons curve in the opposite direction to electrons in a magnetic field.

In 1964, evidence of CP-violating interactions was observed for the first time in particle-physics experiments. The standard model of particle interactions embraced the phenomenon theoretically in the 1970s with the development of the electroweak theory, which unified electromagnetism (the theory of electricity and magnetism) with the weak nuclear force (one of the two forces that hold together the nuclei of atoms, and which is responsible for radioactive beta decay – where nuclei can give off electrons and positrons). The electroweak model has been verified by experiments, yet the degree of CP violation it accounts for is still insufficient to explain the baryon asymmetry – yielding an amount of matter from the Big Bang equivalent to about one galaxy (whereas our Universe is home to some 80 billion). It's hoped new experiments at the Large Hadron Collider particle accelerator, at the CERN laboratory on the Swiss–French border, will offer new insights into CP violation by smashing together subatomic particles to produce showers of smaller particles called 'b quarks', which are a key marker of CP symmetry.

Antimatter power

Some researchers are already looking beyond pure science to practical uses for antimatter. Positrons are already used in PET (positron emission tomography) scans. Here, a patient is injected with a radioactive substance that emits positrons. The substance is embedded in a biological molecule that travels to regions of high-metabolic activity in the body. The positrons it emits then annihilate with nearby electrons, emitting gamma rays and causing the high metabolic regions to glow. This is useful in the treatment heart disease, neurological disorders and cancer. But what about other applications? The quantity of energy per kilogram extracted from antimatter fuel is about 10,000 times what you would expect from nuclear fission. Could antimatter ever be used as a power source? The problem is that antimatter does not occur naturally on Earth. It has to be manufactured in particle accelerators. Currently, making antimatter takes 10 billion times more energy than it releases when it annihilates. Physicists estimate that the world's accelerator labs can make about 10 billionths of a gram of antimatter per year, at a cost of $600,000. At that rate, making 1 g (0.035 oz) of the stuff – equivalent in energy yield to about 10 kg (22 lb) of uranium – would then take several hundred million years and cost about $60 trillion. There is one exciting possibility, though. Antimatter could be mined from space. Violent events in space clash together subatomic particles, some of which break apart to form antimatter. James Bickford, a scientist working at the Draper Laboratory in Massachusetts, has calculated that nearly 4 tonnes of this antimatter drifts into our Solar System every year – enough to meet about two-thirds of the world's annual energy demand. This, and other antimatter formed by cosmic rays, would be attracted by the magnetic fields of the planets – because it's electrically charged – where it would form belts rather like the Van Allen particle belts that surround Earth. A rich mining site would be Jupiter, which has a particularly potent magnetic field – about 14 times stronger than Earth's. Bickford believes a spacecraft equipped with powerful magnetic scoops could be send to retrieve this material.

Storing antimatter

Once gathered, antimatter is stored in a device known as a Penning trap. This is effectively a magnetic bottle, which holds the electrically charged particles in strong electric and magnetic fields to stop them from annihilating with the container walls. For the same reason all the air is evacuated from the space inside. The antimatter in the trap is also cooled using a laser beam to prevent thermal energy freeing it from the grip of the magnets. Laser cooling works by stimulating the antimatter particles to radiate energy.

Antimatter could offer a major boost to space flight itself. The big problem in deep space exploration is the mass of fuel that spacecraft must carry. Antimatter offers a dense, portable source of energy that could conceivably propel a craft around the Solar System or to distant stars on feasible timescales. NASA has calculated that a 100-tonne spacecraft powered by antimatter could reach a speed of 100,000 km/s (62,000 m/s) – fast enough to get to the nearest stars in about 12 years. The spacecraft would use a Penning trap to store its fuel, from where it would be siphoned into a combustion chamber and mixed with particles of ordinary matter. The two groups of particles annihilate spewing fragments out of the back of the rocket engine at high speed, propelling the spacecraft forward. An antimatter engine would be over 2,000 times more efficient than a space shuttle main engine. The amount of antimatter required would still be large, however, compared to what we have available using current techniques. One option a team at NASA's Marshall Space Flight Center has looked at is to use antimatter to boost a fusion-powered nuclear engine (see *How to build an atomic bomb*). They calculated that just 10 micrograms of antimatter would be sufficient to power a spacecraft on a tour of the Solar System. This propulsion system would enable a journey to the outer limits of Solar System in around a year.

Even 10 micrograms of fuel won't come cheap. But as Lawrence Krauss says in *The Physics of Star Trek*: 'It is the ultimate rocket-propulsion technology, and will surely be used if ever we carry rockets to their logical extremes.'

CHAPTER 16

How to see an atom

- Small world
- Electronic eyes
- Quantum tunnelling
- Atomic force microscopes
- Moving atoms

Atoms are some of the tiniest objects in nature, as small as 32 billionths of a millimetre in size. The smallest thing the human eye can see from a distance of 30 cm (12 in) is about 0.1 mm (0.003 in) across. Atoms are more than three million times smaller again. Even the most powerful desktop microscope falls short by a factor of 1,500. However, during the 20th century, physicists have built a range of new high-tech microscopes that are capable of peering all the way down into the atomic realm.

Small world

The word atom was first used nearly 2,500 years ago by the Greek philosopher Democritus to describe the smallest indivisible particles of matter possible. Keep chopping a chunk of matter in half and there comes a point where what you have cannot be divided any further. In 1803, British physicist John Dalton was one of the first scientists to take seriously the idea that matter was made of atoms. It was a great idea, and explained phenomena such as Brownian motion – how dust grains viewed under a microscope seem to be getting knocked this way and that by an invisible force. In Dalton's theory the invisible force came from collisions with atoms. Atoms also laid the foundations for the emerging science of chemistry. The fact that chemical elements are made of indivisible chunks explained why the products of chemical reactions always come in small whole-number proportions – because it's impossible to make a mass of a chemical that weighs less than the mass of one atom. But despite their importance, no one had much clue what these tiny entities looked like. That would start to change in the 19th century as physicists began to piece together a theoretical picture of how each atom is built.

In 1896, British physicist J.J. Thomson was studying cathode rays – beams of particles given off by a negatively charged electric terminal that has been heated up. When he measured the mass of these particles, he found that it was tiny. Thomson had discovered the electron, a minuscule subatomic particle that orbits in the outer reaches of every atom. Later, two other particles would be discovered – the positively charged proton and the electrically neutral neutron, which inhabit the dense core of an atom called the nucleus. At first, it was thought that electrons orbited the nucleus like the planets orbit the Sun. This view changed in the 1920s with the development of quantum mechanics. In this view, electrons are complicated three-dimensional waves that oscillate around the nucleus in accordance with the laws of quantum theory.

Electronic eyes

In the early 1930s, the German physicist Ernst Ruska and electrical engineer Mark Knoll took advantage of the wave nature of electrons to design a new kind of microscope capable of magnifications of one million times, a factor of 500 better than an optical microscope. A few years earlier, the French physicist Louis de Broglie had figured out an equation to calculate the equivalent wavelength of a particle from its momentum – a measure of the impetus of a moving body given by its speed multiplied by its mass. When Ruska and Knoll did this for electrons they found that the de Broglie wavelength of a typical electron is hundreds of thousands of times shorter than the typical wavelength of light. The shorter the wavelength, the greater the level of detail it can resolve, so a microscope that uses an electron beam offers far greater resolution.

The two researchers now had to figure out how to channel and focus beams of electrons in the same way that the lenses and mirrors of an ordinary microscope channel light. The design they finally came up with generates a beam of electrons using a device called an electron gun, a negative electrical terminal that's heated to give the electrons enough thermal energy to escape from the metal. These electrons are then attracted and accelerated by a positively charged grid and focused into a tight beam using a magnet.

The magnet directs the electron beam onto the object to be studied. As the electrons pass through the object, some of them are deflected by its internal structure. When they emerge on the other side, the beam and the image imprinted on it by this structure is magnified by another set of magnetic lenses and then focused onto a fluorescent screen where it can be viewed. Because the electrons are transmitted through the specimen being observed, this technique is known as transmission electron microscopy (TEM). The first TEM was built in 1931. A few years later, Knoll and colleagues built a variant that was able to capture wider-field images. It did this by deflecting the path of the initial electron beam using a set of electrical coils. By varying the current in the coils the beam could be made to scan back and forth across the specimen object – rather like an old tube-based television set that scans an electron beam rapidly back and forth across a phosphor-coated screen to form a picture. This design of electron microscope is normally known as a scanning electron microscope, or SEM for short. The beam in an SEM bounces off the specimen rather than passing through it.

Image resolutions with these techniques can reach scales of 1 nanometre (nm) – a millionth of a millimetre – allowing crisp images of blood cells, micro-organisms, viruses and crystal structures in materials. It still isn't quite good enough to see atoms, but it was a massive step forward.

Quantum tunnelling

A machine to see atoms would finally become reality in 1981 when German physicists Gerd Binnig and Heinrich Rohrer invented the scanning tunnelling microscope, or STM. Rather than illuminating the target specimen with electrons – or any other form of particles or radiation – the STM makes use of a quantum phenomenon known as tunnelling. In classical (non-quantum) physics, two electrical charges of the same sign repel each other. The electric fields set up by the two charges interact with one another to push them apart and this force of repulsion can only be overcome by applying a greater force, i.e. shoving them together very hard. In quantum physics, however, this isn't quite true. Electric charges of the same sign still repel each other but down in the quantum world particles behave like waves – their position in space isn't tied down to a single point but is instead smeared out over a fuzzy patch. This uncertainty in position can bring two charged particles together without the need to apply as much external force to them.

Tunnelling plays a major role in understanding the nuclear fusion reactions at the heart of the Sun. Fusion works by joining together atomic nuclei, resulting in the release of energy. However, nuclei are all positively charged so must be forced together in order to overcome the mutual repulsion they feel. This is achieved by heating the nuclei up, which makes them jiggle around violently and smash together. To collide them with enough force to do this would ordinarily require temperatures well in excess of that found in the Sun's core. Tunnelling solves the problem, reducing the force, and hence the temperature, needed for fusion to take place. STMs take advantage of quantum tunnelling by dragging a tiny probe across the surface of the specimen object being studied. The probe doesn't actually make contact with the surface but remains around two atomic widths away. The tip of the probe itself is extremely sharp – usually just a single atom thick. If a tiny voltage is applied between the tip of the probe and the specimen, a stream of negatively charged electron particles are able to quantum tunnel between the two, setting up an electrical current. The strength of the current varies with the size of the gap between the probe and the surface. A computer alters the probe height to keep the current constant. The changes in height as the probe is dragged across the surface reveal the surface's texture. By scanning the needle back and forth across the surface, a picture of its lumps and bumps can be built up.

This technique can achieve resolutions of up to 0.01 of a nanometre – 1/100,000,000 of a millimetre – good enough to pick out individual atoms, showing them up as peaks that each look rather like a wizard's hat. Technically, these aren't actual images of the atoms but representations of their position based on interactions between their outer cloud of electrons and the STM probe.

Atomic force microscopes

The one problem with STM is that the voltage needed to generate the tunnelling current means that it can only be used to map out the surfaces of electrically conducting materials. That all changed in 1986 with the invention of the atomic force microscope, or AFM. This device is similar to an STM but rather than using a tunnelling current, the tip of the probe itself drags over the surface of the sample object. The probe is mounted on a cantilever arm, which moves up and down in response to the surface contours of the sample. This movement is detected by shining a laser beam on the cantilever – as the cantilever moves, it shifts the angle by which the beam is reflected, which in turn can be measured by a light sensor. The readings are fed to a computer which uses them to build a relief map of the surface and displays it on a screen. AFMs can also be used to measure the mechanical properties of materials, such as their elasticity and roughness. Because there is no requirement for the sample to conduct electricity, AFM is used to study all kinds of materials – from plastics to biological samples.

Moving atoms

In 1989, researchers Donald Eigler and Erhard Schweizer were able to spell out the letters 'IBM' in xenon atoms by nudging the atoms around with an STM probe. The technique works by getting the tip of the probe so close to an atom that the atom actually sticks to it and can be dragged across the surface and deposited where it's needed. This means that an STM probe doubles as an atomic toolkit that can be used to engineer structures on the tiniest scales. Most recently it has been used to alter the bond structure between atoms inside molecules, changing the molecules' chemical properties. In the space of just over 200 years, our understanding of the atomic world has gone from one of complete ignorance to being able to print the fundamental atomic structure of matter on a T-shirt.

CHAPTER 17

How to turn lead into gold

- Ancient alchemy
- Radioactivity
- Structure of the atom
- Nuclear transmutation
- The final piece
- Nuclear waste
- Making gold

The alchemists of old searched in vain for a way to turn humble base metals into gold. Although largely based on pseudoscience and superstition, alchemy laid the foundations for what became the modern science of chemistry. But even that wasn't up to making gold. The solution finally came in the 1930s with the development of nuclear physics: a way to tinker with the very heart of an atom, and ultimately transform one chemical element into another.

Ancient alchemy

Alchemists searched for a legendary substance known as the philosopher's stone, which was believed to have the power to turn ordinary base metals, such as lead, into precious metals, such as gold. But the work of these early alchemists had little to do with any real scientific understanding of the structure of matter. Unfortunately for them, they did find small amounts of gold in the residues of their experiments, which just led them further from the truth. These traces were already in their original samples of ore, and had been separated out by the primitive chemical processes they were using. It turns out that it *is* possible to turn lead into gold, but the secret lay a million miles from the pseudoscientific ramblings of the alchemists. It had to wait for the development of modern physics in the early 20th century.

Radioactivity

In the 1800s, physicists had realized that matter was made of tiny, indivisible chunks known as atoms (see *How to see an atom*). It was discovered in 1896 that some atoms sporadically spit out particles in a process known as radioactivity. This seemed to take one of three different forms – called alpha, beta and gamma. Alpha radiation consisted of positively charged heavy particles; beta radiation consisted of high-speed negatively charged electrons; gamma radiation was essentially electromagnetic radiation – similar to light but with much higher energy.

One such radioactive material is the chemical element thorium. In 1901, British physicist Ernest Rutherford and his colleague Frederick Soddy found that thorium emits alpha particles – and that as it does this the thorium gradually converts itself into a chemical element called radium. One chemical element was being converted into another – exactly what the ancient alchemists had sought to achieve. Soddy called the process 'transmutation' (despite Rutherford's protestations that the name sounded too reminiscent of alchemy). However, quite what was going on – or how the process could be controlled – was still a mystery. That would have to wait for a better understanding of atoms and how they work. But it wasn't far away.

Structure of the atom

Scientists had already figured out that atoms were broadly composed of a piece with positive electrical charge and lots of little pieces with negative electrical charge, called electrons. But how they fitted together was anyone's guess. At the turn of the 20th century, the smart money was on an idea called the 'plum pudding model' – which said that atoms consisted of a positively charged blob (the pudding) with negatively charged electrons embedded within it (the plums). But in 1911, that all went out of the window following a landmark experiment by Ernest Rutherford and his colleagues Hans Geiger and Ernest Marsden.

Rutherford, Geiger and Marsden bombarded a thin sheet of gold foil with particles of alpha radiation. They were expecting the positive charge of the alpha particles to interact with the electrical charges inside the atoms, causing the particle trajectories to be deflected, which they hoped would reveal details about how charge is distributed within each atom. The experiment did just that, though the results proved to be the death knell for the plum pudding model. Rather than seeing each radiation particle deflected by a few degrees, as they were expecting, most of the particles passed straight through the foil unaffected. However, now and again a lone alpha particle bounced back from the foil towards the radiation's source. To Rutherford, the meaning was clear. The positive charge of the atom is concentrated into a tiny volume, so that most of the time the alpha particles passed through undisturbed. But on the rare occasions one strayed near to one of these clumps of positive charge, the particle's own positive charge was repelled, catapulting it back in the direction it had come from. Rutherford's team had discovered the atomic nucleus, a pinprick of positive charge at each atom's core. Indeed, it's so small that if an atom was scaled up to the size of the Albert Hall, the nucleus would only be the size of a pea.

Nuclear transmutation

Rutherford was intrigued by his team's discovery. If the positive charge of the atom was concentrated into a nucleus at the centre then what exactly was this blob of positive charge made of? In 1919, he carried out an experiment that revealed the answer. He fired alpha particles into a cloud of nitrogen gas, and found that some of the particles were absorbed by the nitrogen atoms, which in turn spat out a nucleus of the element hydrogen. Rutherford correctly concluded that the hydrogen nucleus must be a fundamental component of all atomic nuclei. He consequently named it the proton – it has positive electric charge, equal but opposite to that found on the electron. Each atom then has a number of protons in its nucleus together with an equal number of electrons orbiting in a cloud around the outside, making its overall electric charge zero.

The picture of the atom was gradually taking shape. But what distinguishes an atom of one element from an atom of another? A Dutch physicist by the name of Antonius van den Broek already had the answer to that one. A few years before the discovery of the proton, he had noticed that different elements seem to be specified by the charge on their atomic nuclei. Experiments soon confirmed his theory. Rutherford's discovery then made it crystal clear how you go about changing an atom of one element into another – simply alter the number of proton particles that it contains.

It didn't take long for Rutherford to twig that he'd already done this. Each atom of the nitrogen gas in his experiment had seven protons in its nucleus. His alpha particles each contained two protons and each time one of them was absorbed by a nitrogen nucleus it emitted a hydrogen nucleus – which is just one proton. This meant that what was left must be an element with eight (7 + 2 – 1) protons at its centre. This is the gas oxygen. Rutherford had actively created oxygen from nitrogen.

The final piece

There was still a problem with the structure of the nucleus that physicists couldn't quite figure out. Bunching a large number of positively charged protons together in this way should cause the nucleus to fly apart – because two or more electrical charges of the same polarity repel each other. British physicist James Chadwick solved the problem in 1932 when he discovered a second component of the nucleus. Called the neutron, it had previously gone undetected because it carries no electrical charge. But it does have a mass roughly the same as the proton. Chadwick exploited this property to spot neutrons as they knocked protons – which could be detected – from a piece of paraffin wax.

Neutrons nestle inside atomic nuclei between the protons, stopping these positively charged particles from getting so close that their electric repulsion might fling the nucleus apart. With the discovery of neutrons, physicists were in a position to classify each chemical element by a small set of numbers describing its atomic nucleus. The number of protons in the nucleus is called the 'atomic number', denoted by the symbol Z. The total 'atomic mass' of the nucleus is specified by a number A, just given by adding together the number of neutrons and the number of protons. Since protons and neutrons each weigh about the same – 1.6 million-billion-billionths of a gram – the total mass of the nucleus is just this number multiplied by A. Finally, a third number, N, gives the number of neutrons in the nucleus. It is related to the other two by the formula $A = Z + N$. For example, simple carbon has $A = 12$, $N = 6$ and $Z = 6$.

Only Z is needed to specify the particular chemical element; it is possible to vary N (and hence A) while keeping Z the same by adding or subtracting neutrons from the nucleus. Such atoms are known as isotopes. For example, carbon can exist in the form of isotopes that have $N = 7$ and $A = 13$, and $N = 8$ with $A = 14$. Alpha particles also contain two neutrons. So Rutherford's first transmutation experiment in 1919 can be explained by taking nitrogen ($A = 14$, $Z = 7$) and adding an alpha particle ($A = 4$, $Z = 2$) to make hydrogen ($A = 1$, $Z = 1$) plus an isotope of oxygen ($A = 17$, $Z = 8$).

Nuclear waste

In the late 1930s, scientists realized that neutrons don't always help an atom stay together. Very heavy atomic nuclei are actually destabilized by the addition of a neutron, causing the nucleus to split in half to form two lighter elements, and releasing a great deal of energy. The process became known as nuclear fission. It is the basis of modern nuclear power stations, and was the principle underpinning the first nuclear weapons (see *How to build an atomic bomb*).

One of the ongoing problems with nuclear power is disposing of the radioactive waste it produces. This material is filled with elements such as plutonium, neptunium and americium, which remain radioactive for tens of thousands of years, meaning any nuclear waste repository presents a long-term radiation hazard. A solution that scientists are currently investigating is to transmute these radioactive waste materials into more benign elements. They believe this can be done by placing the waste in a specially designed reactor and bombarding it with neutrons to produce new elements that are either non-radioactive or whose radioactivity decays away after a few tens of years. Research projects are underway around the world to make transmutation of nuclear waste a practical, working technology. If they are successful, it could mean clean nuclear energy with fewer dangers attached.

Making gold

But what of our original quest to turn ordinary metals into glittering gold? It is indeed possible. In 1980, US scientist Glenn Seaborg made a small amount of gold (Z = 79) from lead (Z = 82) in a nuclear reactor. Sadly, however, he found the process requires a massive amount of energy – so much so, it cost more per kilogram of gold produced than the metal's market value. In a cruel twist, it is actually much easier to turn gold into lead, which is probably why lead is plentiful while gold is so rare.

Gold is not the only precious metal, however. The metals rhodium and ruthenium are also extremely expensive, and both are produced in nuclear power plants as a bi-product of the energy that's released to generate electricity. And they only remain radioactive for a few years. Japanese scientists now have a plan to begin extracting these metals with a view to selling them in order to offset the large cost of processing nuclear waste. It's not quite the alchemists' dream. But it is a sobering reminder, if one were needed, that claims which sound too good to be true often are just that.

CHAPTER 18

How to build an atomic bomb

- Chain reactions
- Mass deficits
- Fission bombs
- Critical mass
- Fusion bombs
- Nuclear explosions

They remain the most fearsome weapons of war ever created. A single nuclear bomb – built around a piece of radioactive material no bigger than an orange – can flatten an entire city and kill hundreds of thousands of people. A nuclear weapon is frighteningly simple to make. So much so that, today, the fear is that they may be used not only by nation states but by terrorists as well.

Chain reactions

Nuclear weapons are a development that was inspired by a science fiction story. In 1914, H.G. Wells published a novel called *The World Set Free*. It tells the story of scientists who find a way to accelerate the rate of decay of radioactive materials, such as radium, enhancing the amount of radiation they give off. Hungarian physicist Leo Szilard read Wells's novel in the early 1930s. By then, the structure of the atom was well understood: a condensed nucleus containing neutrons and positively charged proton particles, with negatively charged electrons orbiting the outside. Radioactivity was known to be caused by an instability of some atomic nuclei, causing them to spontaneously throw out particles and radiation every now and again. Szilard wondered whether Wells's vision might actually be possible if the particles given out by the decay of one atomic nucleus could stimulate other nearby nuclei to decay in the same way, setting up a chain reaction.

Mass deficits

Meanwhile, other physicists were noticing something strange about atomic nuclei, namely that their masses were less than the mass you get when you add up all their constituent particles. For example, a standard carbon nucleus is made of six protons and six neutrons, but its total mass is found to be less than six times the proton mass plus six times the neutron mass. They called this difference a mass deficit. But what did the mass deficit mean? The answer lay in Albert Einstein's special theory of relativity. Published in 1905, this was a theory describing the motion of objects moving at close to the speed of light. As well as giving us the best description yet of superfast motion, Einstein's theory also threw up what is probably the most famous equation in the whole of physics: $E = mc^2$. It essentially said that mass (*m*) and energy (*E*) are just different aspects of the same fundamental entity, linked together by the speed of light (*c*). Physicists interpreted the mass deficits to mean that some of the mass of the constituent particles in atoms gets turned into energy and released as they come together to form a complete nucleus. This is known as binding energy because it is the energy required to hold the nucleus together. Fire in this amount of energy and you shatter the nucleus apart. Assemble the same nucleus and the binding energy is released as a burst of radiation.

There was another surprise to come when they plotted a graph of the binding energy per particle against the total number of particles in the nucleus. Rather than just a flat line, they found a curve that rose sharply up to a peak then gradually dropped off again as the total number of particles in the nucleus became large. Any nuclear reactions in which the binding energy of the nuclei that come out is more than the binding energy of the nuclei that go in will release energy. A quick look at the graph made it clear to the physicists that they could do this in one of two ways: by joining together two light atomic nuclei to the left of the peak to make a heavier one, a process called fusion; or by splitting apart a heavy nucleus to the right of the peak to make two lighter ones, called fission. This latter possibility was the one physicists chose to pursue first.

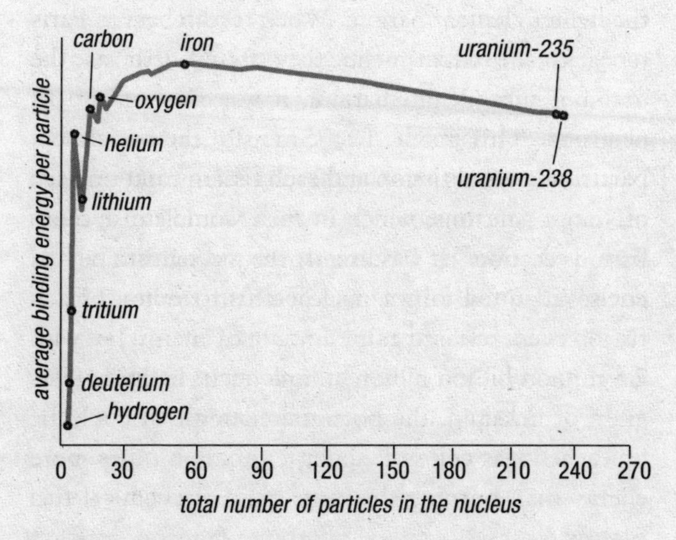

How the binding energy per particle varies with the total number of particles in the nucleus.

Fission bombs

The drop in binding energy towards the right of the graph means that the nuclei become less stable the bigger they get. This led to the notion of a large atomic nucleus as rather like a big, wobbling droplet of water, ready to break apart at the slightest poke or prod. In 1938, German physicists Otto Hahn and Fritz Strassmann decided to do some poking and prodding using neutron particles, firing them at a sample of the heavy isotope uranium-235 (so named because each nucleus contains a total of 235 particles). Sure enough, they found that some of the uranium nuclei broke apart under neutron bombardment to form nuclei of the lighter element barium. When researchers in Paris repeated the experiment, they found that as the uranium turned into barium, it was also giving off neutrons. This made Leo Szilard's ears prick up. Neutrons caused fission and each fission reaction gave off more neutrons, which in turn would cause more fission reactions. It was exactly the mechanism he had envisaged to sustain a nuclear chain reaction. Each fission event released a tiny amount of energy, but with 2.5 million billion billion atomic nuclei in every kilogram of uranium, the potential energy that could be unleashed was colossal: about ten million times more energy than burning the same mass of chemical fuel such as oil.

The discovery led Italian-US physicist Enrico Fermi to fire up the world's first ever nuclear reactor at the University of Chicago in 1942. The reactor consisted of a mass of uranium interspersed with rods made of graphite, which soaks up neutron particles. By inserting or withdrawing the rods, the rate at which the chain reaction proceeded could be fine-tuned: pushing them in slightly soaked up more neutrons, slowing the reaction down, while pulling them out speeded it up. Withdrawing them entirely could lead to a runaway reaction – a nuclear explosion. With war already raging around the world, this was a fact that wouldn't be ignored for long.

Critical mass

The first nuclear weapon to be used in anger was a uranium bomb, dropped on the Japanese city of Hiroshima on the morning of 6 August 1945. The blast, equivalent to the simultaneous detonation of about 15,000 tonnes of conventional TNT explosive, destroyed nearly 70 per cent of the city. It killed 80,000 people instantly and many tens of thousands more from prolonged radiation-related injuries. The Hiroshima bomb, codenamed 'Little Boy', was a so-called 'gun-type' nuclear weapon. Inside the bomb was a long metal tube with a chunk of uranium positioned at each end. Behind one chunk was a conventional high-explosive charge. To set off the bomb, the explosive was detonated, firing one piece of uranium into the other at high speed. The size of the two pieces was carefully calculated so that when they were brought together they exceeded the critical mass needed to trigger a runaway nuclear chain reaction. When the mass of uranium is less than critical, there are not enough fission reactions taking place to generate sufficient neutrons to keep the chain reaction going, and it shuts down. At the critical mass there are just enough neutrons being created to maintain equilibrium, while above it – termed 'supercritical' – there are so many neutrons that the reaction rate increases exponentially. Three days after the bombing of Hiroshima, America dropped a second atomic bomb on Japan, this time over the city of Nagasaki. Still a fission device, this bomb used as fuel the slightly heavier radioactive isotope plutonium-239. And rather than the gun-type detonator of Little Boy this bomb, called 'Fat Man', used chemical explosives surrounding a sphere of plutonium to compress the fuel, squashing it to high density and mimicking the effect of a larger super-critical mass of plutonium. This is known as an 'implosion type' nuclear fission device. Fat Man was equivalent to some 21,000 tonnes of conventional TNT. Despite the higher yield, the hilly terrain disrupted the effects of the blast, making it less devastating than the Hiroshima attack – though it still claimed nearly 40,000 lives. The sphere of plutonium that did all this was about 8 cm (3 in) in diameter.

Fusion bombs

Fusion releases its energy by sticking together lighter atomic nuclei to increase the average binding energy per particle in the nucleus. For example, combining two hydrogen nuclei creates deuterium (or heavy hydrogen – an isotope of hydrogen with an extra neutron in its nucleus) plus a positron (the antimatter counterpart of the electron) plus a lot of energy. However, the atomic nucleus is positively charged, so two nuclei brought close together tend to repel one another. To overcome this repulsion, the nuclei need to be slammed together with considerable force. This is usually achieved by heating the fusion fuel. The kinetic theory developed in the 18th century ascribes the temperature of a gas to the vibration of its atoms and molecules: the higher the temperature, the more vigorously they are jiggling around. Heat a gas sufficiently and the vibrations are forceful enough to surmount the electrical repulsion between nuclei. The temperatures required are colossal – upwards of 8 million °C – which is why fusion reactions are sometimes described as 'thermonuclear'. Once this temperature has been reached and fusion has begun, the energy released sustains the process, leading again to a chain reaction.

Thermonuclear energy is the principal power source of stars, where the core temperature is easily hot enough. In a fusion weapon, such as a hydrogen bomb, these temperatures have to be generated artificially. This is usually achieved using fission to kick start the process. A small implosion-design fission bomb releases X-rays that compress a cylinder of fusion fuel. The cylinder has a plutonium core, which begins a second fission reaction under compression, and this in turn ignites fusion. The yield from a fusion device can be many times higher than that from a fission bomb. Most of today's nuclear bombs are fusion devices.

Nuclear explosions

Nuclear weapons derive their lethality from three key effects: heat, blast and radiation. Heat is given off by the chain reaction, creating a fireball where the temperatures can reach thousands (in the case of a fission bomb) or millions (for a fusion device) of degrees C. Heat has the longest reach of all three effects, lighting fires over a wide area. The blast wave causes most deaths by collapsing buildings and flying debris, and is lethal within a radius about half that affected by heat. Radiation comes in two forms. During the explosion, radiation emitted by the fission process can fatally damage biological cells in the short term, while radioactive ash thrown up by the explosion into the mushroom cloud over the blast site rains back to the ground as fall-out, causing long-term health problems including cancer. Mushroom clouds themselves vary greatly in height. The cloud made by Fat Man over Nagasaki was just a few hundred metres high. By contrast the cloud produced by the largest nuclear weapon ever detonated, the Soviet Tsar Bomba, rose up 64 km (40 miles). Detonated in a test in 1961, this weapon unleashed the same explosive force as 57 million tonnes of TNT.

Nuclear weapons are a modern-day Pandora's Box. Some argue that the spectre of nuclear war has served as a peace-keeping deterrent. That can be little consolation to the children of Hiroshima and Nagasaki, whose forebears were in the wrong place at the wrong time the day the box was opened.

CHAPTER 19

How to harness starlight

- The bright Sun
- Matter waves
- Solar cells
- Solar thermal energy
- Star power

The energy pouring out from our Sun in a single second is enough to meet planet Earth's energy demands for over 800,000 years. Even the small proportion of this energy falling on the planet could sustain us for 1,000 years. It's not surprising then that many scientists believe solar energy to be one of the most promising solutions to the world's energy woes. So how does it work and why aren't we using more of it?

The bright Sun

Greek philosopher Archimedes is said to have realized the power of the Sun over 2,000 years ago, when he used a complex arrangement of mirrors to turn its energy into a heat ray to fend off an invading Roman army. Indeed, the Sun is a veritable powerhouse, kicking out energy at the rate of 400 million billion billion watts. That could run an awful lot of 100 W light bulbs. The Sun derives all this power from nuclear fusion: bonding together atoms of hydrogen in its core to form heavier atomic nuclei and liberate energy in the process (see *How to build an atomic bomb*). According to Einstein's equation $E=mc^2$, which says that to turn a mass (m) into energy (E) you just multiply by the speed of light (c) squared, the Sun's copious power output means it's losing weight at the alarming rate of 4 million tonnes every second. But it is so massive – 2 billion billion billion tonnes – that it could keep up this drastic weight-loss regime for another 15 million billion years, or about a million times the present age of the Universe.

Matter waves

The key breakthrough in the development of solar electricity was the discovery of the photoelectric effect by German physicist Heinrich Hertz in 1887. Hertz observed that certain metals could be made to emit electrons – so generating an electric current – when exposed to electromagnetic radiation. His apparatus consisted of a sparking device that would fire when electrons were released from the metal under exposure to sunlight. But he couldn't work out why the size of the spark decreased when glass was placed in front of the metal, but not when a crystal of quartz was put there instead. It was later realized that glass blocks ultraviolet light, whereas quartz doesn't – meaning that only high-frequency ultraviolet light is able to generate a photoelectric current. Why this should be though was still a mystery. Albert Einstein finally solved the problem in 1905 when he used the idea that light can sometimes behave as particles as well as waves. And that it was collisions with these particles, or 'quanta' of light, like collisions between billiard balls, that were knocking electrons from the metal.

The realization that light waves could be quantized in this way was one of the very first building blocks of quantum theory, a new way of understanding the physics of subatomic particles that was developed primarily during the first half of the 20th century. The notion was first put forward in 1901 by the German physicist Max Planck, who was developing a theory of heat radiation. Planck found that he could explain the characteristics of the electromagnetic radiation given off by a hot object if he assumed that the radiation was emitted in discrete packets – the quanta – each with an energy given by their frequency multiplied by a tiny number, now known as Planck's constant.

Einstein made the jump to interpreting the quanta of light as actual solid particles, named photons. The energy of each of the light quanta, as calculated by Planck, could then be thought of as the kinetic energy of a solid photon. This is the energy that's possessed by any solid body as a result of its motion. And, as anyone who's ever caught a cricket ball or a baseball knows, anything with kinetic energy can deliver a forceful impact. Einstein calculated the minimum kinetic energy of a photon needed to turf an electron from a metal, and then worked backwards using Planck's equation to find out what frequency the corresponding electromagnetic waves would need to have. His findings explained perfectly Hertz's observation that only waves above a particular threshold frequency – corresponding to ultraviolet light – can stimulate photoelectric emission.

Later, the French physicist Louis de Broglie extended the idea that light waves can behave as particles by showing that the reverse is true too – that particles can sometimes be thought of as 'matter waves', with a characteristic wavelength related to their solid-particle energy. This relationship between radiation and matter is known as wave–particle duality. It's a recurring theme in quantum theory. For example, electron particles undergo diffraction when they pass through the gaps in a crystal lattice aperture, while photons of light exert a measurable force as they rain down on a surface.

Solar cells

Modern solar panels work using a variation on the photoelectric effect, called the photovoltaic effect.

This is an electrical phenomenon that takes place in semiconductors – materials that aren't perfect conductors but aren't perfect insulators either. Electrons can flow through a semiconductor to a limited degree, but so can the positively charged 'holes' in the material that the mobile electrons leave behind – and this can lead to electrical materials with interesting properties. They come in two types. Positively charged semiconductors (with an excess of holes) are known as p-type, while those with more negative electrons are called n-type.

In a solar cell, the photovoltaic effect takes place at junctions between these different types of semiconductor materials, known as p-n junctions. The materials in these junctions are both normally based on silicon that has been 'doped' – that is, had impurities added – to skew it towards either n-type or p-type. For instance, doping silicon with phosphorous gives an n-type material; adding boron makes the silicon p-type. A photon that's absorbed by the silicon at the p-n junction in a photovoltaic device won't just generate an electron, as in the photoelectric effect, but an electron-hole pair. The negatively charged electrons then flow towards the positive p-type material in the junction – because opposite charges attract – while, for the same reason, the positive holes flow towards the n-type side of the device. Negative charge flowing, say, to the left and positive charge flowing to the right is the same as a large net flow of negative charge all moving to the left. This is how solar cells generate an electric current. The photovoltaic effect was discovered in the 19th century. However, the first purpose-built solar cell made from p-n semiconductor junction devices wasn't switched on until the 1940s. Early solar cells were woefully inefficient, turning only 1 per cent of the radiant energy falling on them into electricity. Today there exist cells with efficiencies of 30 per cent. What does that mean in terms of their electricity yield? The amount of sunlight arriving at the surface of the Earth is 950 watts per square metre. So at 30 per cent efficiency, a one-metre-square solar panel can generate about 285 W of electricity – sufficient to run a computer or a TV, but not enough to boil a kettle. As solar panel technology has improved, so the price has inevitably dropped too. This has led some enterprising individuals to install solar panels on the roofs of their houses to contribute to their domestic energy requirements. It's estimated that a 2 kW home solar array can provide about half the energy needs of an average family.

Solar thermal energy

Another method to generate energy from the Sun is known as solar thermal energy. This can be used in the home too, to heat water. It works by passing water through a network of pipes – rather like central heating radiators that work in reverse, collecting the heat that falls on them.

Similar designs have been used on a larger scale in warm parts of the world to make electricity. These work rather like Archimedes's heat ray, using arrangements of mirrors to concentrate the Sun's radiant energy to boil water. The steam produced is then used to drive turbines. Some designs have even proposed the use of molten salts rather than water to store the heat energy collected at hundreds of degrees C. One such solar concentrator system in the Nevada Desert is generating electricity at the rate of 64 megawatts (MW) using 760 long, trough-shaped reflectors that focus the Sun's rays onto absorber tubes through which liquid is pumped, heating the liquid to nearly 400°C (750°F). The hot liquid is then passed through a heat exchanger, transferring the heat energy to water, making steam that can spin a turbine. Other solar thermal power plants are located in California's Mojave Desert, Kuraymat in Egypt and Hassi R'Mel in Algeria. It was estimated in 2009 that worldwide solar thermal energy generation totals around 600 MW.

Star power

One of the major applications of solar energy has been in space-flight. Carrying batteries or other kinds of power generation equipment adds a lot of weight to a spacecraft, greatly increasing the fuel load and overall cost of the mission, so being able to tap into the free power from the Sun on the fly can bring huge savings.

The US section alone of the orbiting International Space Station derives power from a total of eight solar array wings, each measuring 35 × 12 m (115 × 40 ft) and incorporating 33,000 photovoltaic solar cells. Together they are capable of generating over 130 kW of electrical power.

Futurologists believe that solar panels in space could one day provide plentiful energy for Earth. Earth's atmosphere blocks about a third of the sunlight reaching the planet, meaning that the solar energy available to a spacecraft in orbit is about 50 per cent more than can be harvested at ground level. And whereas a solar energy plant on Earth must spend 12 hours of every day in darkness, during which time it generates no power, a spacecraft can be positioned so that it's never in the shade.

In 2009, the Japanese Space Agency announced plans to build a 1 billion watt solar power station in Earth orbit. The electricity would be beamed down to the ground using lasers or microwaves, and giant satellite dishes would collect it. The scientists behind the idea believe it could produce electricity at a cost of about 9 cents per kilowatt-hour (the amount of electricity it takes to run a 1 kilowatt appliance for an hour). At the time, the price of electricity in Japan from existing sources was around six times this figure.

Space-based solar power could become even grander in the future. British physicist Freeman Dyson has suggested that an advanced civilization might surround its entire home star with a giant light-gathering device to harness every ounce of energy the star gives out. Such devices have become known as Dyson spheres. They can either be rigid structures or a flotilla of billions of smaller solar power stations forming a vast cloud that envelopes the star.

Russian astronomer Nikolai Kardashev went even further, suggesting that super-advanced races might be able to gather up all the energy released by every star in their home galaxy. This would be a truly mind-boggling amount of energy – with every second's worth being enough to power the present-day civilization on Earth for nearly 3 billion years. There can be no doubt that the power given off by the Sun and other stars is a gold mine of free energy. This just makes it even more baffling why solar energy still only accounts for a few hundredths of a per cent of all the electricity that we consume.

CHAPTER 20

How to visit the tenth dimension

- Curved space
- Kaluza–Klein theory
- String theory
- Compactification
- M-theory
- Escape to hyperspace

Philosophers and mystics have long mused over the possibility that there could be more dimensions to our Universe than the three of space and one of time that we can see. But it took a while for scientists to cotton on and take the idea seriously. Now extra dimensions are a feature of many theories in particle physics and, it's hoped, could soon reveal themselves in experiments.

Curved space

One of the first scientists to consider the possibility of extra dimensions was the German mathematician Bernhard Riemann. In the 1850s, he developed a mathematical formalism to describe curved spaces in any number of dimensions. This was a fantastically powerful tool. Up to that point, scientists had to rely on intuition and geometry, but these approaches only work in one, two or three dimensions. Few humans have the brain power to visualize what a cube looks like in eight dimensions, let alone the skill in geometry to be able to draw one. Riemann's analysis got round that by providing a systematic mathematical framework to study spaces, in principle, in any number of dimensions. It worked by using a matrix of numbers known as a 'tensor' to describe the curvature of space at any point. Generally for an N-dimensional space the tensor would have N^2 components arranged in an $N \times N$ matrix. So in our three spatial dimensions, it's a 3×3 matrix with a total of 9 components.

In the second decade of the 20th century, when Albert Einstein came to develop his general theory of relativity – a theory of gravity based on curvature of space and time – Riemann's equations were invaluable, effectively giving Einstein a ready-made toolkit with which to build his theory. One of the central tenets of relativity was that time is unified with the three dimensions of space to form a continuous four-dimensional fabric that Einstein called 'space–time', described by a 16-component 4×4 tensor.

Kaluza–Klein theory

In the 1920s, two mathematicians, German Theodor

Kaluza and Swede Oscar Klein, took this to the next step. They tried to unify gravity with the force of electromagnetism, the best theory of which had been developed in the 19th century by Scottish physicist James Clerk Maxwell (see *How to cause a blackout*). The electromagnetic field can be specified by one number for every dimension of space plus one for time. Quantities such as this are called vectors. Your position in space is another example – in 3D it takes three numbers (usually labelled x, y and z) to specify where you are. Kaluza and Klein tried to unify electromagnetism with gravity by adding an extra row and an extra column to the 4×4 space–time tensor of Einstein's general relativity and into these blank spaces adding the four components of the vector describing the electromagnetic field. The result was a 5×5 tensor – describing gravity and electromagnetism as curvature of a 5D space–time.

But, as astute readers may have noticed, something was missing. Sixteen components of Einstein's gravitational tensor plus two times the four components in Maxwell's electromagnetic vector only gave 24 components, whereas a 5×5 tensor should have 25. Kaluza and Klein concluded that for their 5D theory to work they needed to introduce an extra field specified by a single number – a so-called 'scalar' quantity. Another example of a scalar is mass. One number – giving the amount in, say, kilograms – is enough to specify how much an object weighs. They interpreted this number as a new field of particles pervading space. Today, these particles are a feature of most theories of fundamental physics that invoke higher dimensions. However, back in the 1920s there was no evidence that such additional particle fields existed, and so the theory was abandoned. Shortly after Kaluza and Klein put forward their theory, a science-fiction writer came up with a name for extra dimensions of space. In his 1934 short story *The Mightiest Machine*, author John Wood Campbell described them using the word 'hyperspace'. And the name has stuck.

String theory

By the 1960s and '70s, the discovery of weird new particles of nature had become a common occurrence. Particle physicists spoke of quarks, gluons, mesons and other subatomic exotica from the particle world. And so Kaluza–Klein theory made something of a comeback. Only it wasn't quite the same theory Kaluza and Klein had originally envisaged. Quantum theory was now well established, and so any fundamental physics theory would need to deal with quantum particles and wavefunctions, and the old theory took account of neither.

The new version of Kaluza–Klein theory would not deal with particles either, preferring to think in terms of pieces of string. The motivation for this was simple:

particle physics didn't work. True, it had provided good descriptions of all the forces of nature, and even successfully unified electromagnetism and the weak force. But that was where the success seemed to end. Attempts to unify the other forces in nature – especially gravity – resulted in divergences, instances where calculations relating to physical quantities give ridiculous infinite answers. Some physicists believed the reason for this stemmed from modelling particles such as quarks and electrons as points of zero size – when in reality they must have some physical extent, or they wouldn't exist. So a few theorists began replacing zero-dimensional point particles in their theories with tiny one-dimensional 'strings' of energy.

As the idea was developed it became clear that a string would behave rather like the string on a guitar, in that waves could exist on it. Waves of different frequencies would produce different notes – and each note would correspond to a particular species of subatomic particle. The calculations also revealed that the strings would be almost infinitesimally tiny – the difference in size between a fundamental string and a proton being equal to the size difference between a proton and the whole Solar System. This naturally meant that testing the theory would be difficult. Physicists were quite capable of building detectors to study protons and electrons but seeing down to the level of strings would be another matter entirely.

This difficulty in testing string theory has brought criticism from many quarters. Proponents say that there are subtle, indirect tests of the theory that can be carried out using the Large Hadron Collider particle accelerator at CERN on the Franco–Swiss border. Like Kaluza–Klein theory before it, string theory predicts the existence of an extra scalar quantity in the equations, corresponding to a new particle of matter. In string theory, this particle is named the 'dilaton' – though it is, as yet, undetected. Also in keeping with the Kaluza–Klein model, string theory requires space–time to have extra dimensions, lots of extra dimensions. The most common versions of string theory operate in a total of 10 space–time dimensions – adding six extra spatial dimensions to the usual three of space plus one of time.

Compactification

If there are all these extra dimensions of space then why don't we see them? The string theorists have an answer to this one too. They say that the dimensions are hidden from view by a process known as compactification, which effectively amounts to rolling them up very tightly. Imagine a sheet of paper. This has two dimensions. Now roll it up as tightly as you can and view it from a distance. The tighter you roll it and the further away you view it from, the more it will look just like a one-dimensional line: one of its dimensions has been compactified. Assuming string theory is correct, you might ask why it is that our Universe has six little spatial dimensions and three big ones. Couldn't we have five big ones and four little ones, or any other combination?

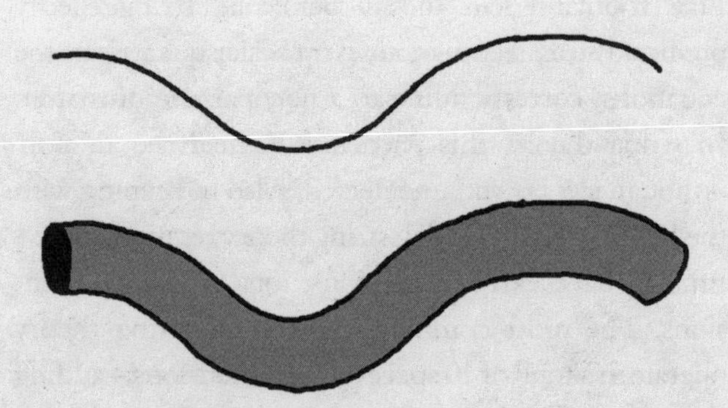

Extra space dimensions are hidden from view because they are compactified. Here, a 2D surface is rolled up tightly so that it looks like a 1D line.

There's a very good reason why our Universe can't have fewer than three large dimensions: because it has life in it. Life requires a flow of energy – living creatures need to eat. And this energy, or food, must come in and pass through some kind of digestive tract, where it is processed and the waste products finally excreted. But in two dimensions such a tract would divide a creature into two. Without the third dimension there can be no extra structure to hold it together. Using the existence of life to make inferences about fundamental physics may seem like a weak argument. But it's actually a surprisingly powerful method of scientific reasoning, called the anthropic principle (see *How to create life*).

The anthropic principle also means it's unlikely the number of large space dimensions could be more than three and still allow us to exist. Physicists have shown that in this case, the orbits of planets around the Solar System and even the motions of electrons around atomic nuclei could become unstable – planets would crash into their parent stars or fly off into space, and matter itself would fall apart. Worse still, pizza wouldn't be very palatable in a large number of dimensions. An ordinary two-dimensional, disc-shaped pizza has a quantity of nice topping proportional to the disk's area – πr^2, where r is the radius of the disk. On the other hand, the same pizza has an amount of rather bland crust that's proportional to the disc's outer circumference – $2\pi r$. That means the ratio of crust to topping in two dimensions is $2\pi r / \pi r^2$, which is just $2/r$. The general form of this formula for an N-dimensional pizza is a little more tricky to calculate, but turns out to be just N/r. So when N is large, your pizza is mostly crust. Yuk.

M-theory

As if 10 dimensions weren't enough, in the mid-1990s US physicist Edward Witten went one better and came up with new theory that uses 11 dimensions. String theory comes in many different versions. What Witten found is that each type of string theory is just a special case of a broader overarching model, which has become known as M-theory.

Rather than modelling subatomic entities as one-dimensional strings, M-theory treats them as two-dimensional membranes. The strings are still there, but they're just 1D slices through these 2D membranes. And the particular orientation of the slicing is what discriminates between each of the string theory variants. But then, of course, space must have one extra dimension over string theory in order to accommodate the extra spatial extent of the membranes – hence 11 dimensions. And just in case you were wondering, no one really knows what the 'M' in M-theory actually stands for. Although suggestions include 'membrane', 'master' – and even that it's an upside-down 'W', for 'Witten'.

Escape to hyperspace

US physicist and author Michio Kaku has suggested that the higher dimensions of string theory could conceivably save humanity from the end of the Universe. Calculations of how the particle physics of the early Universe unfolded in various different string theories show that the expansion of our four-dimensional space–time was in some way coupled to the compactification of the other six dimensions. Kaku thinks that if our Universe ends in a Big Crunch scenario – where, billions of years in the future, the expansion of space eventually turns back on itself and the Universe recollapses into a hot Big Bang-like state (see *How to destroy the Universe*) – then the compactification of the other six dimensions will also reverse, so that they begin to expand. Kaku believes there will come a point, just before our space shrinks too small, where the size of the other dimensions will have become large enough for human travellers to hop across into them. Perhaps luckily for us, this isn't something we're likely to have to worry about any time soon.

CHAPTER 21

How to survive falling into a black hole

- Curved space
- Degeneracy pressure
- How to find a black hole
- Event horizon
- Ripped apart
- Life preserver

There is no escape from a black hole. Once you have crossed its outer boundary, the gravity is so strong that nothing, not even light, can shake its inexorable pull. Falling into a black hole, your body would be stretched out into spaghetti by the immense forces. But now physicists have found a way for an intrepid space traveller to survive the plunge into a black hole's abyss.

Curved space

The modern description of black hole physics had to wait until 1915 and the publication of Albert Einstein's general theory of relativity. This theory replaced Newton's law for strong gravitational fields by describing gravity as curvature of space and time. In Einstein's theory, the gravitational field of an ordinary star forms a bowl-shaped depression in space. Planets orbiting the star can be imagined as rather like marbles rolling around the inside of the bowl. Roll a marble fast enough and it will fly over the edge of the bowl and escape. But as the star gets smaller and denser, so the bowl becomes deeper until – in the case of a black hole – it resembles a long funnel-like tube. Any marbles getting too close are destined to roll in and spiral down the tube, no matter how fast they are moving. General relativity predicted that at the centre of a black hole lies a so-called 'singularity', a point of zero size and infinite density, where the curvature of space and time and the gravitational forces become unboundedly large, crushing anything that encounters the singularity out of existence in a heartbeat.

Degeneracy pressure

Black holes are fascinating theoretical objects but can they really exist in nature? It seems so. Perhaps the most common way they are thought to form is when a massive star reaches the end of its life. The result is a huge outpouring of energy known as a supernova explosion. The explosion crushes the core of the star to high density, increasing its gravity, which then pulls the star in on itself. Ordinary gas pressure is unable to support a ball of material trying to squash itself down

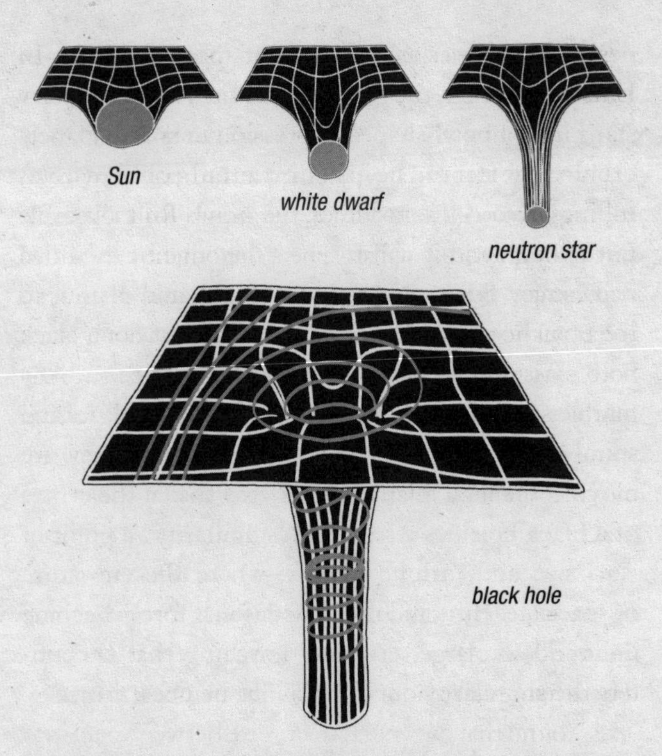

As a gravitational source gets smaller and denser, the gravitational field close to it gets stronger and stronger until a black hole is formed.

in this way. But that in itself isn't enough to make a black hole, as a young Indian astrophysicist proved in the early 1930s.

Subrahmanyan Chandrasekhar worked out that another much stronger force steps in once gas pressure has been overcome. The force resulted from a principle in the emerging field of quantum theory – the laws of physics describing the behaviour of atoms and molecules. One aspect of quantum theory is called the exclusion principle. It was put forward in the 1920s by Austrian physicist Wolfgang Pauli, and in its most elementary form it says that quantum particles don't like to get too close together. Quantum forces literally push the particles apart. The phenomenon is called degeneracy pressure and Chandrasekhar applied it to electrons to show that it's able to support dying stars with masses up to about 1.4 times the mass of our own Sun. A star supported by electron degeneracy pressure is known as a white dwarf – an incredibly dense object packing the mass of an ordinary star into a sphere the size of Earth.

In the late 1930s, US physicists Robert Oppenheimer, George Volkoff and Richard Tolman repeated Chandrasekhar's calculation for larger neutron particles. They found that degeneracy pressure between neutrons can support stellar corpses with masses up to about three times that of the Sun. These objects are known as neutron stars, and are even denser than white dwarfs – squashing the mass of a star into a sphere about the same diameter as a city. If the star going supernova is heavier than about 3 solar masses, then there is no known force in the laws of physics that can support its gravitational collapse – and it must form a black hole.

How to find a black hole

If a black hole is black, and space is black, then how do you know if there's one there? This was the problem facing astronomers trying to investigate what the theoretical astrophysicists were telling them about these weird, otherworldly objects. But in fact astronomers have managed to gather convincing evidence that black holes really do exist. There are several ways they've managed to do this. Sometimes a black hole will exist in a binary system with another, normal star. The two stars orbit around their common centre of gravity so that the presence of the black hole is revealed by its gravitational influence on the motion of its luminous companion star.

Sometimes when a black hole is orbiting in a close binary system, the hole's gravity will tear material from its companion. The material is then sucked into a belt that orbits around the black hole's equator – known as an accretion disc. Material in the disc gradually loses energy and spirals inwards to ultimately be devoured, but as it does so it gets compressed and heats up, emitting X-rays, which can be detected by telescopes back on Earth. Black holes have also been seen lurking at the centres of some galaxies. These so-called supermassive black holes weigh millions of solar masses. Astronomers have inferred their presence by studying stars orbiting close to the centres of these galaxies. The stars are found to be moving so fast and in such tight orbits that the central mass cannot be anything but a black hole.

Our own Milky Way is believed to harbour a black hole in its core weighing 2.6 million times the mass of the Sun. The largest known black hole lies at the centre of the galaxy QJ 287 and weighs 18 billion solar masses. In 1975, Stephen Hawking famously bet the US astrophysicist Kip Thorne that Cygnus X-1, an X-ray source in the constellation of Cygnus, was not a black hole. If Hawking won the bet he would receive a subscription to *Private Eye* magazine; if Thorne won he would receive a subscription to *Penthouse*. In 1990, Hawking conceded, and most physicists now accept that black holes really do exist. So what might it be like to fall into one?

Event horizon

The outer surface of a black hole is known as its event horizon. It's not a solid surface, but a sphere traced out by the distance from the singularity at the hole's core at which the strength of gravity is too strong for light to escape from it. In 1916, German physicist Karl Schwarzchild calculated the radius of the event horizon around a black hole. For a body of mass m, the Schwarzschild radius is $2Gm/c^2$, where G is Newton's gravitational constant (1/15,000,000,000) and c is the speed of light (300,000,000 m/s). Any object can become a black hole if it is squashed small enough. For example, the Schwarzchild radius for the Sun is about 3 km (1.8 miles). If Earth were turned into a black hole it would have an event horizon with a radius of 9 mm (0.35 in).

An astronaut falling towards a black hole event horizon would notice the gravitational field gradually start to increase as she drew close. Another effect would soon kick in too. A bizarre consequence of general relativity makes time in a gravitational field appear to run more slowly, as measured by a distant observer watching the whole process through a telescope. It's a similar effect to the time dilation experienced by observers travelling at close to light speed. This is caused by light having to spend energy climbing out of the hole's gravitational field. It's called the gravitational redshift effect and has been verified experimentally. On the event horizon itself the magnitude of this effect becomes infinite and time there appears to freeze. Gravitational redshift also means that the light from the astronaut is gradually stretched out to lower wavelengths the further she falls towards the black hole until it's eventually shifted outside the visible spectrum and she fades serenely from view. From the astronaut's own point of view it's a far less gentle ride.

Ripped apart

As the astronaut starts to approach the hole, she starts to see stars behind it distorted by the intense gravitational field. Starlight that would normally go nowhere near the astronaut's eyes is hooked around the hole by its gravity so that to the astronaut it appears as if they are viewing space through a fisheye lens. As she gets closer, the effect intensifies. At 1.5 times the Schwarzschild radius from the black hole's centre, the astronaut encounters the 'photon sphere'. At this distance, the gravity is strong enough that light can orbit in a circle around the hole. Firing her jet pack for a moment to hover on the photon sphere, the astronaut looks left and right and finds that she can see all the way around the hole so her gaze falls on the back of her own head.

Falling feet-first towards the event horizon, she notices the difference in the force tugging her feet and her head get larger. The gravity field is so intense that even though her feet are only a metre and a half closer to the black hole than her head, the extra gravitational force they feel is colossal. This force begins to stretch her body out. At the same time, the force squashes her body laterally across the shoulders. The nearer she gets to the singularity, the more pronounced the effect becomes until ultimately her head and her feet are pulled far apart and her body is stretched out into a long, thin strand.

Life preserver

Ordinarily gravitational forces inside a black hole would tear a traveller apart in about 0.1 of a second. But two US physicists – Richard Gott and Deborah Freedman – have come up with a way that the traveller could buy themselves a little extra time. They've calculated that a massive ring of material around the traveller's waist could cancel out some of the force they experience. When they say 'massive' they really do mean massive – approximately the weight of a large asteroid. The ring's gravity would act to pull the traveller's feet and head back together and to counteract the squashing forces pressing in on them. Far from the black hole the ring would be about the size of one of the rings of Saturn, but as the traveller got closer it would shrink down, strengthening its gravitational effect to counteract the hole's increasing gravity.

On its own, the ring will only give you an extra tenth of a second, doubling the time you could survive. But this might just be enough to save your life. The secret lies in the physics of rotation. Karl Schwarzschild's early black hole studies just looked at holes created by a stationary mass of material. But in 1963, a New Zealand-born mathematician called Roy Kerr solved Einstein's equations for the black hole created by a mass that's spinning. The so-called Kerr solution made a fascinating prediction. Whereas the singularity at the centre of a stationary black hole is a point, which you can't avoid running into once you've crossed the event horizon, the singularity inside a rotating black hole takes the form of a ring. If the black hole is big enough – and with the help of a Gott–Freedman life preserver – it could just be possible for a human traveller to pass straight through this ring. Calculations suggest that the traveller would emerge on the other side of the ring into a new region of space – though quite where this region would be physicists can't yet say for sure. Some speculate it may be a distant region of our own Universe; others suggest it could be a new universe entirely.

For this reason, many scientists have interpreted Kerr black holes as wormholes – tunnels through space and time (see *How to travel through time*). Virtually all objects in space have some degree of rotation – think of planets, the Solar System, and the galaxy – meaning that Kerr black holes may well be the norm rather than the exception. If that's the case, then falling into a black hole might not always be the death sentence it's often made out to be. Far from it – when it comes to bridging the gap between our Universe and others, it could be the safest way to travel.

CHAPTER 22

How to see the other side of the Universe

- Eddington's eclipse
- Gravitational lensing
- The ultimate telescope
- Microlensing
- Multiply-connected universes

Today, telescopes on Earth, and in orbit around it, are so powerful they could resolve the gap between two car headlamps on the Moon. But sometimes even these mighty instruments need a helping hand. A phenomenon called gravitational lensing – where the light from distant galaxies is focused by gravity – is boosting the magnification of terrestrial telescopes to reveal objects tens of billions of light years away, at the very edge of space.

Eddington's eclipse

In 1919, the British astronomer Sir Arthur Eddington led an expedition to the African island of Principe to view a total eclipse of the Sun. But this was no mere sightseeing trip. Eddington was there to test one of the most radical theories in physics ever put forward. Called the general theory of relativity, and authored by one Albert Einstein, the theory supposed that the gravitational interaction between objects is caused by the deformations their masses cause to the structure of space and time. If this was correct then anything passing through space should experience the distortion – including beams of light. On the other hand, the existing theory of gravity, put forward by Isaac Newton, made no such claim over light.

Eddington proposed to test Einstein's theory by measuring the degree to which a light ray gets bent by the gravity of the Sun. The biggest light-bending effect would occur where the Sun's gravitational field is strongest: when a light ray just skims across its surface. But there was a problem. The Sun is extremely bright and so any such rays, say from distant stars, will be swamped by the solar glare. Except, that is, during a total eclipse. When this happens the Moon passes in front of the Sun. By sheer coincidence, the angular size of the Sun and Moon are identical. Even though the Moon is physically smaller than the Sun, it's much closer to us – by just the right amount so that its disc at totality exactly obscures the disc of the Sun. Eddington planned to take advantage of this darkness during the total eclipse to measure the effect of the Sun's gravity on the apparent positions of stars nearby in the sky. His observations were almost thwarted by bad weather, but

Eddington was successful in taking a number of pictures of the eclipse, and these confirmed the light-bending effect – showing the degree of bending to be in good agreement with the predictions of general relativity.

Gravitational lensing

If light bends around the Sun then it should also be bent by other massive objects in space, such as galaxies. In 1924, Russian physicist Orest Chwolson published an article outlining how the same effect should take place on vast cosmic scales so that the light from very distant galaxies gets distorted by the gravity of bodies along the line of sight to Earth. Chwolson pointed out that if the intervening body is exactly on the line joining the distant galaxy to the observer, the distant galaxy will appear not as a point but as a ring.

To see why, imagine for a second that space is like a two-dimensional sheet of paper. Now draw a straight line on the paper and mark along it three points: planet Earth, the distant galaxy and, exactly between the two, the intervening mass doing the lensing. Rather than travelling directly along the straight line, the gravitational distortion makes the light from the galaxy travel to Earth along two curved arcs, one above the line and one below it. A two-dimensional astronomer on Earth would see two images of the galaxy either side of its true position on the sky, one arriving along each arc.

To see what happens in three dimensions, simply rotate the two-dimensional diagram around the axis formed by the original straight line. The arcs along which light travels now sweep out a shape resembling a rugby ball, and the two images that are formed in two dimensions now trace out a ring. The light from the galaxy is curved around the intervening mass in much the same way as light passing through a lens. And for this reason the process has come to be known as gravitational lensing. If the mass doing the lensing is slightly off axis, then the perfect ring breaks up into a number of images of the distant galaxy, stretched out into arc-like segments. However, Chwolson's research was not a particularly in-depth analysis. Albert Einstein picked up the baton in 1936, writing one of the definitive papers on gravitational lensing, and building an extensive mathematical framework by which astronomers could calculate the degree of bending produced as a function of distance and the masses involved. The bright ring produced by a perfectly lensed galaxy is now usually known as an 'Einstein ring'. The first real one was found in 1998 by a team using NASA's Hubble Space Telescope.

The ultimate telescope

The first such gravitational lens was discovered in 1979, when astronomers spotted a double image of a faraway type of galaxy known as a quasar. Quasars are fiercely bright galaxies seen at the edge of the visible

Universe. Hundreds of thousands of them are now known, though none is closer than about 3 billion light years to Earth. Because of the look-back time – the time taken for their light to reach us – this means we're seeing quasars as they were at least 3 billion years ago, suggesting that these objects were perhaps a phase in the evolution of young galaxies that has now finished. Quasars were discovered in 1963 by Dutch–US astronomer Maarten Schmidt. The name is a contraction of 'quasi-stellar object', which they were given because it wasn't immediately recognized that they were galaxies. Their great distance from Earth had made them appear as points of light in the sky, so that they looked more like stars (hence 'stellar objects').

The double quasar that marked the discovery of gravitational lensing was 8.7 billion light years away. If that seems a lot, the farthest quasar ever seen (as of early 2010) is an astonishing 28 billion light years away. Gravitational lenses don't just distort the light from faraway galaxies and quasars but they magnify it too. Light rays which ordinarily would have meandered away into empty space are captured by the intervening cluster's gravity and hooked around onto an intercept course with Earth. More light from the quasar reaches Earth than would do were it not lensed. This enables us to see quasars that are further away than we would normally be able to see.

But this advantage comes at a price. For any particular quasar, the chances of there being a massive galaxy or cluster of galaxies directly on or close enough to our line of sight for lensing to happen are slim, and so the number of lensed quasars observed is very small compared to the total number of quasars that are out there. For many other distant quasars, the gravity of other galaxies and galaxy clusters must pull light away from a path that would ordinarily intercept Earth, rendering these objects too faint for us to detect.

Microlensing

It's not just objects on the other side of the Universe that astronomers can study using gravitational lensing. The magnification effect that it causes makes it an excellent way for spotting the passage of objects within our galaxy that would otherwise be invisible to telescopic eyes. The object itself is not magnified, but rather it causes magnification of the light from background stars as it passes by.

One class of faint object this technique is used to search for are so-called brown dwarf stars. These are failed stars, which were never massive enough to create the temperature and pressure at their cores needed to spark up nuclear fusion reactions (see *How to create life*). Physically, these objects are thought to resemble the planet Jupiter – vast balls of gas, but not emitting any light or other radiation. Because they are dark, brown dwarfs are extremely hard for astronomers to spot in the blackness of space. Until, that is, they happen to wander in front of a background star. When that happens the mass of the brown dwarf gravitationally lenses the light from the star, causing it to momentarily grow brighter. The mass of a brown dwarf is typically less than 7 per cent the mass of the Sun or, equivalently, about 75 Jupiter masses. This means that the lensing effect is small, which is why it is known as microlensing.

Studying how the light from a lensing event grows and fades can reveal information about the object doing the lensing, such as its size, speed and distance. In 1998, researchers at the University of Sussex even suggested that microlensing might be one way to look for wormholes. These are tunnels through space and time, whose existence is predicted by general relativity (see *How to travel through time*). Wormholes require a special kind of material to hold them open, which has negative mass and so generates a kind of repulsive gravity. This deflects the light rays around the wormhole in a markedly different way from how the rays are normally bent around, say, a brown dwarf, fanning them out rather than focusing them inwards, to make a double-peaked lensing event. The first peak reaches maximum brightness slowly and then darkens rapidly, to be followed a few years later by a second peak that brightens rapidly and then fades out gradually. Astronomers could in principle look for this behaviour in the light from microlensing events to discover wormholes – objects which no one has yet seen in the real world.

Multiply-connected universes

According to one slightly bizarre theory of the Universe, if you use a gravitational lens (or any other kind of powerful telescope) to peer into the outer reaches of space, you could actually end up peering in on somewhere a little closer to home.

In a so-called multiply-connected universe, space is wrapped around on itself so that light rays travelling far enough in one direction ultimately arrive back where they started. The possibility arises because Einstein's theory of general relativity says nothing about a property of space called topology. Broadly speaking, this determines the overall shape of space and how different regions are connected to one another. For example, a flat sheet of paper has different topology from the surface of a sphere – because on a sphere you can travel all the way around without ever arriving at the edge. Glue opposite sides of the sheet of paper together and you get a different kind of topology again, resembling a doughnut. Now you can travel on a loop either around the outside of the doughnut or through the hole in the middle and arrive back where you started. However, it is possible to set a course that spirals around the ring and never quite brings you back to your starting point.

There are far more complicated kinds of topology than spheres and doughnuts. In 2003, a French-led team suggested that our Universe could have a weird topology based on a 12-sided dodecahedron, arranged so that if you exit on one face you re-enter through the face opposite. They claimed to have found tentative evidence for their theory in the pattern left behind in the cosmic microwave background radiation – the microwave echo of the Big Bang fireball in which our Universe began.

The claim is controversial. However, space probes expected to launch over the coming decades to measure the microwave background in unprecedented detail could provide the conclusive evidence one way or the other. After all of the astronomers' efforts to gaze ever further out into the blackness of space, they could ultimately end up zooming in on the backs of their own heads.

CHAPTER 23

How to recreate the Big Bang

- The Big Bang
- The microwave background
- Particle cosmology
- Particle accelerators
- Hunt for the Higgs
- The end of the world?

Some fear it will destroy the world. Scientists, on the other hand, say it will utterly revolutionize our view of the cosmos. Either way, the Large Hadron Collider particle accelerator, on the southern border of France, is the most complex machine ever built by human beings. It will accelerate subatomic particles to 99.99 per cent of light speed and slam them together, generating temperatures over 100,000 times hotter than the Sun's core in an effort to reconstruct the fiery conditions of the Big Bang in which our Universe was born.

The Big Bang

Somewhere around 13.7 billion years ago, something incredible happened. Out of the void of total and complete nothingness – no matter, no radiation, not even any space or time – our Universe popped into existence in a quantum event known as the Big Bang. The cosmos was born into a state known as a gravitational singularity – a point of zero size and infinite density, temperature and pressure. By rights this primeval maelstrom should have collapsed back and disappeared again as suddenly as it appeared. But the matter that had just been brought into existence had other ideas. About one hundred-million-billion-billion-billionth of a second after its creation, the stuff of the embryonic universe underwent a phase transition – a wholesale shift in its properties, rather like the conversion of steam into water as it cools. Unlike water, though, this change was to have dramatic consequences. It filled space with a kind of anti-gravitating material – rather like the 'dark energy' that's thought to pervade the Universe today (see *How to destroy the Universe*). This caused the Universe to expand stupendously fast, increasing in size by a factor of 10^{26} – a 1 followed by 26 zeroes. To put that in perspective, if at the start of this phase of cosmic 'inflation', as cosmologists call it, the Universe was about the size of a tennis ball, then by the end of inflation that tennis ball would have grown to be about a billion light years across: about the distance to the furthest galaxies observable in the Universe today.

Inflation not only dug the Universe out of a gravitational hole by stopping it collapsing on itself, it also planted the seeds from which galaxies would later grow. As is often the case, it all comes down to quantum physics. Ordinary empty space is filled with virtual quantum particles constantly popping in and out of existence. But as the first quantum physicists discovered in the early years of the 20th century, subatomic particles can equally well be thought of as waves. In the embryonic Universe, these waves introduced tiny irregularities in the density of matter from point to point across space. Before they could pop out of existence again, as is usually the case with such quantum fluctuations, inflation blasted them up from the quantum realm and stretched them out, quite literally, to astronomical proportions. It was the gravity of these now cosmic-sized density fluctuations that sucked in the material from which galaxies grew, and within each galaxy then formed stars, planets and – on at least one planet orbiting an average yellow star in a fairly typical galaxy – there emerged life.

The microwave background

The history of the Universe after the era of inflation, about 10^{-36} seconds (one billion-billion-billion-billionth of a second) after the Big Bang is reasonably well understood. The radiation from the Big Bang fireball was stretched out and diluted by the expansion of space to form a kind of relic echo of the Big Bang, called the cosmic microwave background radiation (CMB). The CMB dates from when the Universe was just 300,000 years old. Before this time matter existed as a sea of positively charged protons and negatively charged electrons. These particles scattered the radiation from the Big Bang this way and that. However, at 300,000 years the temperature of the Universe dropped sufficiently for protons and electrons to combine into electrically neutral atoms. At this point interactions between radiation and matter ceased, and the radiation was free to stream out through space as the CMB. This ghostly microwave glow still pervades space today, and numerous spacecraft have been sent into orbit to study it. Space missions are the best way to investigate the CMB because a great deal of the microwave radiation from space is masked by microwave emission from water in Earth's atmosphere. A small quantity of the CMB does make it through: about one per cent of the static on an untuned TV is, in fact, the echo of creation. Spacecraft have measured the minuscule fluctuations and ripples in the CMB and from it extracted a picture of the post-Big Bang universe. But the physics of the Universe leading up to inflation is something of an undiscovered country. It is believed that the Big Bang was a quantum event – analogous to the creation of virtual particles in empty space – which caused space and time to appear where before there was nothing at all.

Particle cosmology

Only a theory of 'quantum gravity' can address this, the ultimate question of creation, and at present such a theory seems a long way away. Other forces in nature have been quantized successfully. However, Einstein's theory of general relativity – our best theory of gravity – seems exceptionally difficult to cast in quantum language, being riddled with so-called divergences where the predicted values of physical quantities become unphysically infinite. String theory and M-theory are contenders to furnish us with a theory of quantum gravity, but they're yet to deliver on this promise.

Up to about 10^{-43} of a second after the Big Bang, quantum gravity was wrapped up with quantized versions of the other three forces of nature – electromagnetism, and the strong and weak nuclear forces – to form a kind of all-encompassing superforce that physicists have modestly termed 'the theory of everything'. The end of the Universe's quantum gravity era was marked by a phase transition where gravity split away to become a separate force in its own right, leaving the other three bundled together as a so-called 'grand unified theory'. A number of different models for grand unified theories exist, but at present these are very poorly tested.

Another phase transition brought about the end of the grand unified phase of cosmic history – at around 10^{-36}

seconds after the Big Bang. This is the same event at which inflation is believed to have taken place. The grand unified phase transition saw the strong force – the force responsible for binding together the nuclei of atoms – peel away from the rest of the group. Now just electromagnetism and the weak nuclear force (the force responsible for radioactive decay of nuclei) remained unified, as the so-called electroweak theory. This theory is well-established, having been confirmed by experiments. Finally, the electroweak symmetry itself broke apart at around a trillionth (10^{-12}) of a second after the Big Bang to leave the four forces of nature as the separate and distinct entities that we see today.

Particle accelerators

The Universe is the ultimate physics laboratory, providing a stage on which to test out theories of high-energy particle physics through the effects they've had on cosmic structure and the microwave background radiation. But the particle physics processes that happened before inflation are beyond the reach of even the most powerful telescopes. And this is where particle accelerators come in. These are colossal machines that accelerate beams of subatomic particles to extraordinarily high speeds and then smash them together. The idea is to recreate the superheated conditions of the Big Bang in order to study the processes that governed this step in the Universe's birth and development. Particles crash together and are split into their constituent components, such as quarks (each proton and neutron particle is made up of three of these smaller quarks), and the behaviour of these fragments can then be studied to unravel the laws governing their behaviour.

The world's first particle accelerator was switched on at the University of California, Berkeley, in 1931. It had a collision energy of 1 mega electronvolt (MeV). An electronvolt (eV) is a unit of energy defined as the total kinetic energy gained by an electron when it's accelerated through an electric potential difference of 1 volt. There are now 26,000 accelerators in operation around the world. The biggest and most powerful is the Large Hadron Collider (LHC) at the CERN research centre on the border between France and Switzerland. Its maximum collision energy will be 574 tera electronvolts (TeV) per particle – 574 million times more powerful than the 1931 machine.

Particle accelerators work using a tube dotted with a sequence of electrified sections connected to an alternating voltage. Electrically charged particles – such as electrons, protons, or whole atomic nuclei – are placed at one end of the tube. The first tube section is then charged with the opposite polarity to the particles so that they are attracted towards it. As they pass by, the polarity is reversed so that they are repelled onwards and accelerated further. This process is repeated all the way along the accelerator tube so that fast moving particles emerge at the other end. Modern accelerators also incorporate magnets to curl the particles around into a giant ring. This means that the particles can be looped round over and over, gathering more speed with every circuit. The LHC at CERN is a vast underground ring accelerator 27 km (17 miles) in circumference. At full power, particles in the beam complete 11,245 circuits of the ring per second.

Hunt for the Higgs

One of the principal reasons the LHC's sprawling complex of magnets, computers, scientific experiments and heavy duty engineering was put together was to look for a single subatomic particle of nature: the Higgs boson. It's the only missing particle in what has become known as the standard model of particle physics, which explains all of the interactions between the four forces of nature as they stand today as well as the electroweak unified theory. Also known as the 'God particle', the Higgs was proposed in 1964 by the British physicist Peter Higgs as a way for all the other particles of nature to have acquired their different masses. His idea was that space is filled with a sea of Higgs bosons, which cluster around other particles, so making those particles massive. Different kinds of particles interact with the Higgs to differing degrees and this is why the particle types all have different masses – while some remain massless.

Finding the Higgs has had to wait for a particle accelerator with the power of the LHC. That's because the higher the mass of a subatomic particle is, the higher the collision energy required to create it experimentally. The precise mass of the Higgs isn't known, but the bulk of the range in which that mass is thought to lie was above what could be reached with the collision energies of previous accelerator machines. Not everyone is convinced the Higgs exists. Cambridge physicist Stephen Hawking has placed a bet that the LHC won't find it. If he wins, this would mean a substantial rewrite of the laws of particle physics. Then again, Hawking has placed a number of scientific wagers in the past and his hit rate has been less than impressive.

The end of the world?

The LHC will crash together subatomic particles so hard that the density of matter generated in the collisions may well be high enough to form microscopic black holes. This has prompted fears from the public that one of these black holes could 'escape from the lab' and devour Earth. One group in Hawaii even went so far as to file a lawsuit against CERN to try to prevent them from switching the accelerator on. CERN's designers, and other physicists, insist that the accelerator is sáfe. They say that any miniature black holes formed would rapidly evaporate away by Hawking radiation (see *How to make energy from nothing*). Also, the high-energy cosmic rays from deep space that regularly pummel Earth pack much more energy that the LHC collisions. If there really were a danger, the planet would have been destroyed long ago.

CHAPTER 24

How to make the loudest sound on Earth

- Sound waves
- Wave properties
- The decibel scale
- Ultrasound and infrasound
- The brown note
- The Doppler effect
- Shock waves
- Louder than bombs

Bang your fist on a desk and you create sound – a mechanical vibration that travels through solids, liquids and gases. We use it to communicate, as a tool, and as a weapon. It's used extensively in the animal kingdom, while sounds generated in the rest of the natural world are some of the loudest on record. The eruption of the volcano Krakatoa in 1883, for example, was heard nearly 5,000 km (3,000 miles) away. But there was another cataclysm during Earth's prehistory that was even louder still.

Sound waves

Sound is essentially a pressure wave travelling through matter. All matter has pressure inside it due to the thermal motion of its atoms and molecules. The pressure inside a sealed container of gas is caused as the molecules of the gas beat against the container walls. Heat the gas up and their beating becomes more vigorous, and the pressure rises. Pressure is defined as the force per unit area acting on a surface, and it's usually measured in Pascals, after the 17th-century French mathematician Blaise Pascal, who carried out much of the initial research in this area.

Sound is a momentary increase in this pressure. Strike the skin of a drum and the downward movement of the drum skin compresses the air that it's moving into. With nothing to contain it, this compression spreads out through the air as a wave. Close to the drum the compression is greater and the sound is thus louder; as you get further away the wave has spread out, so the degree of compression is less and the sound is quieter. Unlike water waves, which are transverse waves (that is, the displacement caused by the wave is at right angles to its direction of motion), sound waves are longitudinal (the displacement is parallel to the direction of motion, rather like waves on a stretched spring).

Wave properties

Physicists describe sound waves, and all other kinds of waves, using four principal quantities. The first is the speed of the waves. This is determined by the properties of the medium that the sound is travelling through. In particular, it is fixed by the density (with the sound speed decreasing as the density gets higher) and the stiffness (increasing with the degree to which the medium resists being compressed). The second quantity is the frequency of the sound. This is simply the number of wave crests passing a fixed reference point every second. It is measured in cycles per second (also known as hertz, after German physicist Heinrich Hertz). The frequency of a sound is determined by its source – strike a drum once per second and it will give out sound pulses at a frequency of 1 Hz. Rig the drum up to a machine that can strike it 100 times per second and you get sound with a frequency of 100 Hz. We hear the frequency of a sound as its pitch – for example, a middle octave A note is characterized by a frequency of 440 Hz.

The next property is wavelength. This is just the distance between two successive wave crests. It is given by dividing a sound wave's speed by its frequency. And finally, there's the amplitude of the wave. This is the size of the displacement it produces as it passes. Thinking for a moment in terms of water waves, the amplitude is simply the height of the wave crest. Of course, sound is a longitudinal wave so the amplitude is more like the amount of compression of the medium that the wave causes as it travels by. We hear the amplitude as the loudness, or volume of the sound.

The decibel scale

Scientists measure the loudness of a sound on the decibel scale. This is a direct measure of the sound wave's amplitude. It is gauged logarithmically, scaled so that a decibel increase of 40 corresponds to an increase in the amplitude of the sound wave by a factor of 100. Thus, 60 corresponds to 1,000; 80 to 10,000; and so on. Leaves rustling in a breeze come in at about 10 decibels, rain falling is about 50, a loud concert is about 115 and a shotgun blast at close range 170. The threshold of pain begins at about 130 dB.

The idea that sound can inflict pain has led researchers to develop sonic weapons. Since 2004, the US military has been fitting its warships with devices called LRADs (long range acoustic devices), which can project a beam of sound with a loudness of up to 150 dB over a distance of up to 600 m (2,000 ft). The weapons work using a dish about 1 m (3 ft) across, which keeps the sound in a tightly collimated beam. This narrow beam allows it to send intense sound over a considerable range.

LRADs are non-lethal, but the pain they induce is meant to repel attackers in small boats, which might be hard to hit using conventional gunfire. Since 2004, the devices have also been mounted on civilian and commercial vessels to repel pirate attacks. Police forces around the world are now employing them for use in crowd control. Some acoustics researchers even believe that lethal sonic weaponry could be a real possibility in the years and decades to come. Any sound above about 150 dB is considered seriously hazardous to health, and there are already devices used in research applications that can generate sound waves with amplitudes of up to 180 dB.

Ultrasound and infrasound

At present these devices only produce ultrasound, which is sound with a frequency above the upper limit of human hearing (about 20,000 Hz). This makes the devices safe to work with. Ultrasound is now being investigated as a replacement for the motor bearings in computer hard drives, where the hard disc that stores the data inside the computer is levitated on a cushion of rapidly vibrating air. Ultrasound is also used in some electronics factories, where a beam of high-intensity ultrasound can generate enough heat to melt solder and thus fix electrical components in place, and do so with extremely high precision.

Lower-intensity ultrasound is used for medical scans – especially of unborn babies, where sound waves offer a less harmful method of scanning compared to, say, X-rays. The technique works by beaming high-frequency sound waves into the region to be scanned and measuring the time it takes for the waves to bounce back from each point to build up a picture of the tissue within.

At the other end of the scale is infrasound – sound with a frequency below the lower limit of human hearing (which is about 20 Hz). Experiments have shown that infrasound can induce feelings of anxiety, fear and sadness – and for this reason it has been cited as the explanation for many reports of paranormal activity. Studies by the late British researcher Vic Tandy have found high levels of infrasound in sites reputed to be haunted – including the Edinburgh Vaults, a system of tunnels that are said to be haunted.

The brown note

Infrasound has also been at the centre of one of the great urban myths of acoustics research. That is, the existence of something known simply as the 'brown note'. This is meant to be a frequency of sound, between 5 and 9 Hz, that corresponds to the resonant frequency of the human colon. So the story goes, if you subject someone to a loud enough blast of the brown note it will force them to evacuate their bowels immediately. Attempts to prove this idea – including one high-profile investigation on the TV show *MythBusters* – have failed to turn up any credible evidence. And for this reason the brown note is generally regarded as an old wives' tale.

However, the physical phenomenon that the brown note myth is based on, – resonance – most certainly exists. If you strike a tuning fork on a surface and let it ring, the note you hear is the fork's so-called resonant frequency. Next put the tuning fork on a speaker and gradually increase the frequency of the sound played through the speaker. The amplitude of the fork's vibrations will gradually increase until – at the resonant frequency – they become especially strong, reaching a peak, before subsiding again as the frequency grows larger. Resonance is an especially important consideration in earthquake zones, where tremors at the resonant frequency of a building can cause major damage.

The Doppler effect

Resonance is a way by which the amplitude of a sound wave can change dramatically. But the frequency can alter too – when the source of the sound and the person observing it are moving relative to one another. This is called the Doppler effect, and it's the reason why the pitch of an emergency vehicle siren sounds slightly higher when the vehicle is moving towards you and lower when it's moving away from you.

Imagine a sound source moving towards you. After each wave crest is emitted the source moves a short distance so that when the next crest is given off the distance between the two is shorter than it would have been were the source stationary. Because the frequency of the sound is given by the wave speed (which is unchanged no matter how fast the source is moving) divided by the wavelength, the decrease in wavelength translates into an increase in frequency – the pitch goes up. Similarly, when the person listening to the sound is moving towards the source, their motion increases the number of wave crests passing them every second, thus increasing the frequency of the sound. For the same reason, the sound's frequency decreases when they travel away from the source.

The Doppler effect, named after the 19th-century Austrian physicist Christian Doppler who discovered it, has a host of applications. These include mapping blood flow in ultrasound scans by the shift in frequency of an ultrasound wave bouncing off a moving mass of blood. And in submarine warfare, the shift in the frequency of the sonar pings that are sent out by a submarine and then reflected back by surface warships reveals the speed at which the warships are travelling.

Shock waves

It is possible for objects to travel faster than the speed of sound in a medium. In ordinary air, this is 343 m/s, or 1,235 km/h (767 mph). When anything, say an aircraft, travels faster than this, the Doppler effect is the least concern. Sound trying to travel away from the front of the aircraft literally cannot move fast enough to get away from it. As a result the sound waves pile up to form a shock wave – a discontinuous jump in pressure, which anyone nearby hears as a 'sonic boom'.

Aircraft aren't the only things to create shock waves. Nuclear explosions generate supersonic blast waves that propagate outwards from the detonation point. Thunder is the sonic boom created as a lightning bolt instantaneously heats the air around it to tens of thousands of degrees.

Louder than bombs

The sonic boom from a nuclear explosion can reach levels of well over 200 decibels – though if you're that close to an atomic detonation, a bit of noise will probably be the least of your worries. A space shuttle launch is fairly deafening too, coming in at as much as 170 dB. And even an accelerating dragster manages a fairly ear-popping 160 dB – easily enough to cause permanent hearing damage if you stand too close or don't wear proper protection.

It turns out that nature can generate some fairly cacophonous noises ofits own. The loudest animal on the planet is also the largest: the blue whale. Its mating call can be heard from hundreds of kilometres away, and reaches a peak level of nearly 190 dB. The loudest natural sound on record was the eruption of the Indonesian volcano Krakatoa in 1883. It was heard 4,800 km (3,000 miles) away in Alice Springs in central Australia. If you had been standing right next to it, the noise level would have been somewhere in the vicinity of 280 dB.

This still wasn't the loudest noise to have ever rocked the planet during the tenure of human beings. That honour goes to the eruption of the supervolcano under Lake Toba, in Sumatra, 74,000 years ago. Probably the largest volcanic eruption in the last 25 million years, it flung some 2,800 km^3 (670 cu miles) of material into the air and devastated 20,000 km^2 (8,000 sq miles) of land. Estimates suggest this mighty volcanic outburst may have clocked in at a staggering 320 dB – louder than pretty much anything else that's graced the planet since.

CHAPTER 25

How to destroy the Universe

- The cosmological constant
- Hubble's law
- Weighing the Universe
- Dark matter
- Last entertainment
- The Big Rip

Nearly 14 billion years ago, our Universe was born in a superheated fireball known as the Big Bang. Space, time, matter and radiation all popped into existence, expanded and cooled to form the stars and galaxies we see adorning the heavens on a clear night. But how might this grand cosmic tableau come to an end? Scientists foresee several possible scenarios for the Universe's ultimate demise. None of them will be particularly pleasant for anyone around to see it.

The cosmological constant

The fate of the Universe rests entirely on the answer to a single question: how much matter does it contain? The Universe is in essence one great big gravitational system. Of the four forces of nature – gravity, electromagnetism, and the strong and weak nuclear forces – only gravity is important for describing the large-scale motion of the galaxies and the fabric of space and time itself. Our best model of gravity is Albert Einstein's theory of general relativity. The theory ascribes the gravitational motions of planets, stars and galaxies to curvature of the space and time on which they rest. This curvature is fixed by the mass of these objects – plus all the other mass and energy in its various forms that the Universe contains.

Einstein himself was the first person to try to apply general relativity to the whole Universe. On balance, he probably wished he hadn't. This was in 1917, just two years after he had first formulated the theory. He immediately ran into a problem. The equations of relativity were suggesting to him that space cannot sit still. It should either be contracting, falling in on itself under its own gravity; or expanding, with galaxies flying apart from one another fast enough to beat their mutual inward pull.

The best astronomical observations of the time suggested that the Universe was neither expanding nor contracting, but was instead static. As a result, a baffled Einstein did what many other physicists have done before and altered his theory to fit the observations. He introduced a fudge factor, which he called the 'cosmological constant', which effectively made gravity repulsive rather than attractive at long range. This repulsion would counteract the short-range attractive force, and so hold the Universe in a static configuration. Physically, the cosmological constant amounted to positing the existence of energy locked away in empty space. It seemed like a great idea for about 10 years. Then, in 1929, US astronomer Edwin Hubble and his assistant Milton Humason made a discovery that would change everything.

Hubble's law

Hubble and Humason had noticed the work of another US astronomer Vesto Slipher working at the Lowell Observatory in Arizona. Slipher had carried out a study of the light reaching Earth from faraway galaxies, and had noticed something odd. When the light from a galaxy is broken up into its spectrum to reveal the brightness at each particular wavelength, it should show a pattern of bright and dark lines. This is caused as the atoms in the stars that a galaxy is made of absorb and emit light at particular wavelengths, determined by the energy levels of their electrons (see *How to be everywhere at once*). Slipher saw this characteristic pattern in the light from his galaxies, but it wasn't where it should be. The sequence of lines had been shifted to longer wavelengths, in other words towards the red end of the spectrum. The light from the galaxies was thus said to be 'redshifted'.

We see the same redshifting effect in sound waves here on Earth, caused by the Doppler effect (see *How to make the loudest sound on Earth*). When a sound source is moving away from you, the sound's wavelength gets stretched out to become longer (making its pitch, or frequency, lower).

The implication was clear: these galaxies are all moving away from us. Hubble and Humason decided to investigate further. They studied Slipher's galaxies looking for a certain type of star known as a Cepheid. These are variable stars, the brightness of which changes over time and with a period that's directly linked to their average brightness. So measuring the periods of Cepheids in the galaxies told them how intrinsically bright these stars were. Then, by measuring the apparent brightness of the stars through a telescope, the astronomers were able to calculate by how much the light had dimmed with distance and so how far away each one of them was. Next they plotted the newfound distance to each galaxy against the degree of redshift in its light. When they did this a distinct pattern emerged: the redshift increased with distance. This is exactly what you would expect in an expanding Universe. A good analogy is the surface of a balloon. Blow up a balloon slightly and then draw dots on its surface with a marker pen. Now blow the balloon up fully. As it inflates all the dots move away from every other dot, and the rate at which any two dots recede from each other is proportional to their separation. Applied to the Universe, this relationship between expansion speed and cosmic distance is known as Hubble's law.

Weighing the Universe

Einstein was left with egg on his face following this discovery. It turned out that he had modified a perfectly good theory. If he'd played his cards right, he could have used general relativity to predict cosmic expansion ahead of it being observed, and so moved even higher up the echelons of physics greatness. But if Einstein thought he had problems, that was nothing compared to what this meant for the Universe. With the idea of a safe, static cosmos gone for good the question was: where might this cosmic expansion lead in the long term? To answer this, astrophysicists needed to determine whether the expansion would continue forever, or whether gravity might one day put the brakes on and bring space caving back in on itself. To do that they would need to measure how much gravity-generating mass and energy there is in our Universe.

Dark matter

It soon became clear that even this wasn't going to be a simple task. You might think it simply boils down to having a look through a telescope and totting up how much bright matter you can see in the Universe. But astronomers soon realized that most of the matter in the Universe cannot be seen. In fact, what we see turns out to be just a tiny fraction of what's actually out there. Astronomers know this because mass makes its presence felt through the force of gravity. In the 1970s, astronomers noticed that spiral galaxies – galaxies like our own Milky Way that consist of a flattened, rotating disc of gas and dust – are rotating too fast to be held together by the gravity produced by their visible material alone. The laws of gravity predict that the rotation speed should gradually diminish to zero as you move outwards from the galaxy's centre. But astronomers found that once clear of the bulge at a galaxy's centre, the rotation speed was more or less constant. The only way around this seemed to be that the galaxy was embedded within a halo of invisible material. Further evidence had already been seen in galaxy clusters, where the masses of the galaxies seemed insufficient to gravitationally bind them together into a group given the speed of each galaxy's own random motions. So it was that astronomers got their first whiff of what has since become known as 'dark matter'.

In 1998, a group of astronomers led by American Saul Perlmutter were making measurements of a particular type of supernova explosion – outbursts marking the deaths of massive stars – in distant galaxies. These explosions have a very well-defined brightness, meaning that, as with the Cepheid variable stars, they can be used to gauge distance. The data gathered by Perlmutter's team seemed to show that not only is the Universe expanding, but that the expansion is accelerating. But what was causing this acceleration? The only thing known to have this kind of anti-gravitational effect is the energy locked away in empty space. It was Albert Einstein's cosmological constant back from the grave. US cosmologist Michael Turner dubbed this material 'dark energy', and the name has stuck. Now all the astronomers had to do was to calculate the exact amounts of ordinary matter, dark matter and dark energy that our Universe is made from.

Last entertainment

Speculation was growing over what this all might mean for the fate of the Universe. There were two principal scenarios. If the total density of matter in the Universe is greater than a number called the 'critical density' (an average of about five hydrogen atoms per cubic metre), the cosmic expansion will eventually come to a halt and gradually reverse. Slowly galaxy redshifts will start to become blueshifts as gravity hauls the Universe back in on itself (as galaxies begin rushing together, the frequency of their light is shifted towards the blue end of the electromagnetic spectrum). The cosmic history that astrophysicists slaved away to piece together over the course of the 20th century will now be reversed. Galaxies will crash together and merge. Space gets hotter until the superheated conditions of the Big Bang are recreated. The particle processes of the early Universe are undone, as the four major forces of nature merge back into one. And then in an instant the Universe winks out, crushed into a superdense state called a 'gravitational singularity' from where it's gone just as quickly as it first appeared. Scientists have dubbed this the Big Crunch.

The other possibility is that, if the density of space is greater than or equal to the critical density, or if the acceleration caused by dark energy is large enough, the Universe will continue to expand for ever. Rather than burning itself out in a cataclysmic Big Crunch, it fades away gradually. Slowly the last stars burn out, using up the last of the hydrogen and helium nuclear fuel, making it impossible for a new generation to form. With the stars gone, all that's left are neutron stars, white dwarfs – and black holes, which gradually swallow the other objects up, along with any last smatterings of gas and dust. After a near-eternal 10^{100} (1 with 100 zeroes after it) billion years, these black holes themselves have evaporated away by Hawking radiation, all the protons and neutrons that make up ordinary matter have decayed and the dregs of particles and radiation that remain have been stretched apart and diluted by the cosmic expansion to virtually nothing. At this point the

Universe is truly dead. This bleak scenario is known as the Heat Death.

The Big Rip

Detailed observations made by spacecraft in recent years suggest that our Universe is made up of about 4 per cent ordinary matter, 22 per cent dark matter and 74 per cent dark energy. And oddly enough, this all adds up to pretty much bang on the critical density. With the extra outward shove provided by the huge dark energy component it seems there can be little doubt we are heading for a Heat Death. Or are we? In 2003, US cosmologist Robert Caldwell put forward a third alternative for cosmic Armageddon – known ominously as the Big Rip. Caldwell wondered what might happen if the dark energy filling the Universe took a particularly extreme form – known as 'phantom energy'. He calculated that in a phantom-energy-dominated Universe, the rate of cosmic expansion will accelerate to become so large it will tear apart galaxies, stars, planets, people and eventually subatomic particles as well. Cosmological observations aren't yet good enough to say whether phantom energy rules, though even if it does things aren't set to get ugly for another 22 billion years. Plenty of time for some bright physicist to come up with a plan.

CHAPTER 26

How to travel faster than light

- Space speed record
- Ion engines
- Solar sails
- Ultimate speed limit
- Warp drive
- Fantasy fuel?

To say light is rather quick on its feet is perhaps the biggest understatement in physics. At a speed of 300 million m/s, it could run the hundred metre race in about three millionths of Usain Bolt's world record time of 9.578 seconds. Could human beings ever hope to move that quickly, or maybe even quicker? And, if so, how?

Space speed record

The *Apollo 10* spacecraft fell back to Earth in May 1969 having just completed the dress rehearsal for the Moon landings two months later. The re-entry capsule accelerated to 11,107 m/s. That's about 40,000 km/h (25,000 mph). It is the greatest speed ever achieved by a manned spacecraft, and the crew – Thomas Stafford,

John Young and Eugene Cernan – remain the fastest men in history. Yet *Apollo 10*'s speed was just 0.004 per cent of light speed. Light moves unbelievably quickly, covering the distance from London to New York in just 0.02 of a second.

It is certainly possible to travel faster than *Apollo 10* using a rocket, but the amount of fuel needed quickly becomes colossal. Early 20th-century Russian space flight visionary Konstantin Tsiolkovsky worked out an equation for the fuel requirements of rocket journeys into space. It showed that the mass of fuel the rocket has to carry grows exponentially with the speed you want it to reach. You might have expected the relationship between fuel and speed to be 'linear' – in other words, if accelerating the rocket by 100 m/s requires 1,000 kg of fuel then a linear relationship would mean that to reach 200 m/s will require just another 1,000 kg, so a total of 2,000 kg. But exponential growth means that you need many times this amount of fuel.

Ion engines

One way to get around the enormous fuel demands is to run your spacecraft on something with a little more oomph. Tsiolkovsky's equations revealed that the maximum speed attainable by a rocket is proportional to the speed at which it spits out its exhaust. Use a new fuel that spits out exhaust gases five times faster and your spacecraft can go five times as quickly as it would have. It turns out that ordinary rockets, which burn liquid oxygen and liquid hydrogen to release the chemical energy stored in these fuels, produce a relatively slow exhaust. For a Space Shuttle main engine – one of the best rockets that engineers have designed to date – the hot gases move at around 4,400 m/s (14,400 ft/s). But that's nothing compared to the speeds that can be achieved using a new kind of spacecraft engine, known as an ion drive. The latest experimental ion drives can generate exhaust speeds of over 200,000 m/s (670,000 ft/s) – nearly 50 times better than the Space Shuttle.

Standard rocket engines burn fuel in a confined space: the engine's combustion chamber. As the fuel burns it expands, raising the pressure in the combustion chamber and causing the hot gases to rush out at high speed. Ion drives work very differently. Rather than burning the fuel, they accelerate each particle using an electric field. This is possible because the fuel particles are electrically charged ions. Normally fuel is made of atoms that are electrically neutral. An atom consists of a nucleus within which are proton particles, which carry positive electrical charge. Around the nucleus orbit negatively charged electrons. Normally there are an equal number of electrons and protons, giving the atom a net charge of zero. An ion, however, has slightly more or slightly fewer electrons, making its overall charge non-zero. This means it can be accelerated by an electric field, in much the same way that electric current is made to flow through a wire by a battery.

First put forward in 1906 by US rocketry pioneer Robert Goddard, ion drives are now a tried and tested technology. In 1998, NASA launched its *Deep Space 1* robotic probe, which was powered by an ion drive using xenon gas as fuel. Although a relatively low-powered design, with an exhaust speed of slightly over 30,000 m/s (100,000 ft/s), the mission was a resounding success. Many of the world's other space agencies have now flown ion drives of their own – and the power of these devices has been steadily increasing. In 2006, the European Space Agency (ESA) carried out a test of a new ion thruster with an exhaust speed of 210,000 m/s (690,000 ft/s). The drawback of ion drives is that they can only accelerate a small mass of fuel at a time, making the acceleration they deliver very gradual. This low thrust means ion engines are little use for launching spacecraft from Earth's surface, where a high impetus over a short space of time is needed.

However, the slow rate of fuel consumption is balanced by the fact that ion drive engines can run continuously for a very long time: days, weeks, even months. Once in the zero-gravity environment of outer space, a spacecraft equipped with ESA's new ion drive and carrying a fuel load making up 90 per cent of its total mass (the typical proportion for a chemical rocket) could reach a speed of nearly 700,000 m/s (2,300,000 ft/s). That's a vast improvement on *Apollo 10*, but still just 0.2 per cent light speed.

Solar sails

Even faster speeds could be made possible using a novel kind of spacecraft propulsion that does away with fuel entirely. Known as a solar sail, it uses a vast sheet of silvered material to literally hitch a ride on the light streaming outwards from the Sun. In our everyday experience light behaves like a wave, but it can also be thought of as a hail of tiny solid particles, known as photons. Just as the particles of air in a strong sea breeze impart some of their momentum to the sails of a yacht, a solar sail gets a steady push from the photons that make up sunlight, causing it to gather speed.

As with ion drives, the acceleration generated by a solar sail is gradual but continuous, and able to produce very high velocities over long enough time scales. According to some estimates by scientists at ESA, the top end could be as quick as 25 per cent light speed: 75 million m/s (250 million ft/s).

Ultimate speed limit

So how do we go faster still? This is where the laws of physics start to make life difficult. Albert Einstein's special theory of relativity describes the dynamics of rapidly moving objects. He formulated the theory in 1905 in response to weird discrepancies that had arisen between the existing laws of motion and the seminal theory of electricity and magnetism developed by Scottish physicist James Clerk Maxwell in the late 19th century.

The trouble seemed to lie in the way the standard laws of dynamics handle the relative motion of objects travelling at, or close to, the speed of light. The relative motion of two bodies moving towards one another is normally given by just adding their speeds together. So if two cars are driving towards each other at 80 km/h (50 mph) then the speed of one car relative to the other is 160 km/h (100 mph). Applying the same rationale to two oncoming beams of light would mean they were converging with a relative velocity of twice light speed. But this contradicted Maxwell's theory, which seemed to be saying that light – and all other electromagnetic phenomena – should look the same to an observer no matter how fast they are moving.

Einstein constructed a theory of relative motion where the speed of light stays the same for all observers. This is the key postulate of special relativity. However, keeping the speed of light constant in all frames of reference leads to distortions of space and time that bring about some very weird consequences indeed. The first is called length contraction. As a moving body approaches light speed its length in the direction parallel to the motion, as measured by a stationary observer, shrinks. This isn't an illusion – the actual physical length gets shorter.

Stranger still, time slows down for the moving observer, an effect called time dilation. For example, each second ticked off on the watch of someone on a spacecraft moving at 0.99 light speed takes just over 7 seconds as measured on the watch of a stationary observer. You can get an idea of why time dilation happens by remembering Einstein's postulate that the speed of light must stay constant no matter how fast you move. Stretching out the length of each second means that a light beam, as seen from the spacecraft, seems to cover more ground each second, so that its speed relative to the accelerating craft always remains the same.

But there was a third revelation, one with massive consequences for spacecraft engineers on a quest to go faster. And it was this: the faster a spacecraft travels, the more energy it takes to make it go any faster. Like length contraction and time dilation, the effect is unnoticeable at everyday speeds. But as a rocketship approaches the speed of light, the energy required to accelerate it further grows and grows, becoming infinite at the light barrier itself. The physical interpretation of this was plain and simple: according to special relativity travelling at light speed or faster is impossible.

Warp drive

As far as high-speed travel is concerned, the special theory of relativity seemed to be nature's way of saying 'thus far and no further'. Then Einstein came up with a new theory and everything changed again. He realized that special relativity was not compatible with gravity. The best theory of gravity at the time was the universal theory of gravitation, formulated by Isaac Newton in 1687. This had gravity as a force that travels at infinite speed so it is felt by all objects simultaneously, which is manifestly at odds with Einstein's special relativity, where nothing can travel faster than light.

Einstein's solution was ingenious. He imagined space and time as a stage on which physics is played out. In special relativity the stage is flat. Einstein built gravity into the theory by allowing it to be curved by the matter it contains. Announced to the world in 1915, the new theory was called general relativity and it was soon verified to high accuracy by astronomical measurements of the Solar System. While special relativity said that nothing can move across the space–time stage at faster than light, general relativity placed no such constraints on the movement of the stage itself.

Nearly 80 years later, in 1994, Mexican physicist Miguel Alcubierre published a theoretical paper detailing how general relativity could be used to build a 'warp drive' – a way to travel faster than light by bending space and time, named in honour of the science fiction TV show *Star Trek*. Alcubierre imagined a hypothetical rocket ship. His idea was to arrange matter around the ship in just the right way that its gravity would cause space to rapidly expand behind it while the space in front would contract at the exact same rate. The effect would be to rapidly increase the distance between the ship and its starting point and at the same time shrink the distance between the ship and its destination – sweeping the piece of space containing the ship to the destination arbitrarily fast.

Fantasy fuel?

The only problem was that when Alcubierre solved Einstein's equations of general relativity to find out what properties the matter would need to have, he found that it was very bizarre indeed. He needed a kind of material that has negative pressure and mass, and is so weird that even physicists describe it as 'exotic matter' (see *How to travel through time*). Tiny amounts of exotic matter have been made experimentally. However, calculations have suggested that building a working warp drive would demand a quantity of exotic matter equal to a third of the mass of the Sun. So while scientists may have designed the definitive vehicle to break the light barrier, it seems that – just like the space rockets of today – the vast quantities of fuel required may ultimately turn out to be its downfall.

CHAPTER 27

How to travel through time

- Time dilation
- Back to the past
- Exotic matter
- Gateways to the past
- Paradox lost
- Time lock

All of us are shuffling forwards through time at the sedentary rate of 60 seconds every minute. But could we ever supercharge our temporal travels and jump ahead into the future, or even reverse the flow of time to travel backwards and explore the past? There is nothing in the laws of physics as they stand to rule time travel out. Scientists have devised a host of theoretical schemes by which it could happen, and many of them believe these ideas will one day become reality.

Time dilation

It is already possible to travel forwards into the future. In 1905, Albert Einstein put forward his special theory of relativity, a new way of looking at the motion of bodies moving at close to the speed of light. One of its central predictions was a phenomenon called time dilation, which essentially says that a moving clock ticks slower than a clock at rest. Close to light speed, time practically stands still. This applies to all clocks – mechanical, digital and biological. It means that an astronaut who climbs aboard a spacecraft and accelerates up to near the speed of light might only age by a second for every year that passes back on Earth as his clocks are slowed down. When he returns after a year, as measured in his own time, he finds that more than 31 million years have passed on his home planet. Time dilation is more than just a theory – it's a real effect that's been measured in experiments.

Back to the past

The trouble with setting off into the future like this is that at present no one quite knows how you would get back to your own time. Travel into the past is an altogether harder problem that has not yet been solved in practice. In theory, however, scientists have put forward a number of schemes that could earn them an awful lot in lottery winnings, and certainly enough money to buy tickets to see Chopin play live.

Most of these ideas revolve around Einstein's general theory of relativity, published in 1915. Whereas special relativity described space and time as 'flat' and took no account of the force of gravity, general relativity adds gravity into the mix by curving space and time. The crux of Einstein's new theory was a set of equations linking the curvature of space and time to the matter they contain. Just a year after Einstein published general relativity, an Austrian scientist called Ludwig Flamm came up with a mathematical solution of its equations describing what is known as a wormhole – a tunnel through the fabric of the Universe providing a shortcut linking regions vastly separated in space. The term 'wormhole' wasn't actually coined until years later by the US physicist John Archibald Wheeler, who likened them to the hole munched through an apple by a worm. The distance from one side of the apple to the other via the wormhole is shorter than the distance around the apple's surface. That wormholes could be useful for time travel wasn't realized until much later. In 1986, a team of researchers led by US physicist Professor Kip Thorne calculated how Einstein's old idea of time dilation could be used to turn a wormhole into a tunnel not just through space, but back through time as well.

The basic idea was to have one mouth of the wormhole on Earth and to put the other on a spacecraft and fly it off at close to the speed of light for a year. Just like the fictional astronaut from earlier, the wormhole mouth on the spacecraft travels forwards through time by 31 million years. The wormhole mouth itself only ages by a year – because of time dilation – and, crucially, it remains connected to the other mouth on Earth, which has also aged by just a year. Here comes the clever bit. Anyone in the year 31 million AD who now jumps into the wormhole mouth on the spacecraft will emerge from the mouth that remained on Earth just a year after the spacecraft set off on its journey – they have travelled 31 million years into the past.

Exotic matter

Travelling back through time using a wormhole looks great on paper. The trouble starts when you begin working out the engineering details needed to turn the theory into reality. Einstein's equations of general relativity reveal the kind of matter that's needed to create and hold open a wormhole big enough for a person to squeeze through. It is the same weird stuff that we encountered in the last chapter, so-called exotic matter. Exotic matter has negative pressure inside it. Try to blow up a balloon with the stuff and the balloon actually deflates. The energy associated with this negative pressure generates a kind of negative, or repulsive gravity, and this is what holds the wormhole tunnel open. Exotic matter isn't the kind of material that you typically find lying around in large quantities. But minute amounts of it have been observed in a phenomenon known as the Casimir effect. This effect causes two metal plates placed just a few billionths of a metre apart in a vacuum to feel a force pulling them together. The force is caused by the negative pressure of exotic matter being created between the plates.

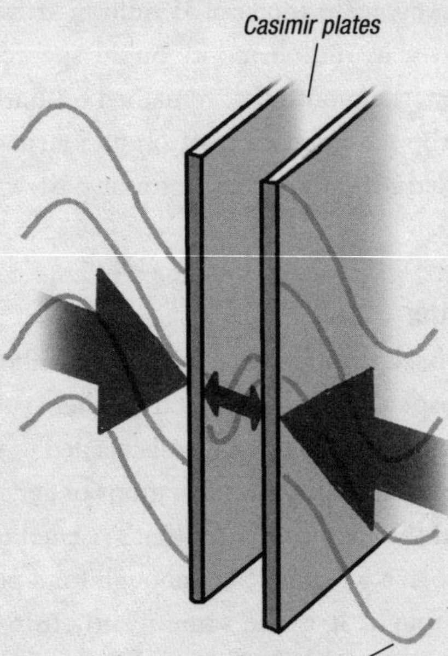

In the Casimir effect, fewer waves are permitted between the plates than outside them, creating a negative pressure.

The Casmir effect happens because empty space isn't really empty. It is actually a bubbling mass of so-called virtual particles – subatomic particles of matter that pop in and out of existence on extremely short time-scales in accordance with the uncertainty principle of quantum theory (see *How to be everywhere at once*). Another aspect of quantum theory called wave–

particle duality (see *How to harness starlight*) says that these particles can equally well be thought of as waves. Between the plates, these waves are a bit like vibrating guitar strings – where the only vibrations allowed are those for which the length of the string is a whole number of half wavelengths.

In the Casimir effect, this means that only waves for which the gap between the plates is a whole number of half wavelengths can exist. Outside the plates, however, all waves are allowed. Converting back to particle language, this means that there are fewer particles rattling about between the plates than there are outside them. In other words, the pressure between the plates is lower. If the space outside is a zero-pressure vacuum, then the pressure inside must be less than zero – there is negative pressure.

The Casimir effect is named after Dutch scientist Hendrik Casimir, who first predicted it in 1948. It was verified experimentally in 1997 by physicist Steve Lamoreaux in New Mexico. The amount of exotic matter produced in the Casimir effect is tiny, around a billion-billion-billionth of a gram. By comparison, sustaining a man-sized wormhole tunnel demands a quantity of the stuff roughly equal to the mass of Jupiter.

Gateways to the past

The other drawback to using wormholes to visit the past is that it's impossible to go back to an era before the time machine was created. If your spaceship with a wormhole in the cargo hold left Earth travelling at near light speed today, you could never hope to use the resulting time machine to visit the Cretaceous Period, the D-Day landings, or even yesterday afternoon.

As well as the technical difficulties in building a time machine, some physicists and logicians have raised objections on the grounds of the causal inconsistencies that it might give rise to. For example, the 'granny paradox' asks what would happen if you went back in time and killed your maternal grandmother before she gave birth to your mother. In that case, you would never have been born, and so could never have gone back in time and killed her, in which case you would have been born, and so on. Another is the 'free lunch paradox', where a generous time traveller takes copies of all seven Harry Potter novels back in time and makes a gift of them to the young, impoverished J.K. Rowling, who promptly copies them, publishes them and never looks back. In this scenario, where exactly did the creative spark for the precocious young wizard originate?

Paradox lost

Time travel paradoxes seem like show stoppers, yet canny scientists have realized that there is a way to travel into the past without disrupting the order of cause and effect. One possible way out of time travel paradoxes could come from the many worlds interpretation of quantum physics (see *How to live forever*). In a nutshell, this says that our universe is just one of many in a sprawling structure known as the multiverse – and that every time our universe is confronted by multiple possibilities it splits into a number of new universes, where each possibility is actually played out. If you believe the many worlds interpretation then time travel is automatically paradox free. That's because travelling back in time takes you into the past of a different universe from the one that you've come from. Assassinating your grandmother in this new universe has no bearing on the version of her in the universe where you were born. Similarly sending all your Harry Potter novels back in time will only place them in the hands of another J.K. Rowling living in a parallel universe. The version in your home universe will still have to earn her fame and fortune the hard way. Another possibility is an idea known as self-consistency. This says that if someone or something travels back through time there will always be at least one self-consistent sequence of events – and that nature favours such an outcome.

Time lock

Nevertheless, some physicists remain staunchly averse to the idea of journeying back through time. At the forefront of these temporal sceptics is British physicist and mathematician Stephen Hawking. He finds the notion so abhorrent that he's put forward what he calls the 'chronology protection conjecture' – a hypothetical mechanism to rule out time travel, either by destroying a time machine as it tries to form, or by destroying anyone or anything that tries to use one. Hawking has yet to find a solid mechanism within the laws of physics by which the conjecture could be implemented, though the virtual particles responsible for the Casimir effect offer one possibility, becoming magnified to destructive energies as they loop through the time machine over and over again.

If Hawking is right then time travel in our Universe will never be possible. Then again, if he's wrong (and he has been before), we're simply waiting for technology to catch up with the science. It wouldn't be the first time. Five hundred years ago, Leonardo da Vinci produced a design for a glider. It was never built, but modern reconstructions of the model have confirmed that had he possessed the advanced materials and construction techniques available to aeronautical engineers today, it would have flown. How ironic that time travel may also be a case of the right idea, but at the wrong time.

CHAPTER 28

How to contact aliens

- Are we alone?
- Where to look?
- How many ETIs?
- From the ashes
- DIY SETI
- The 'Wow!' signal
- Mr President ...

Are we alone in the Universe? Few areas of science capture the imagination in quite the same way as the search for extraterrestrial intelligence, also known as SETI, a project by astronomers to detect signals broadcast by alien civilizations. A new network of radio telescopes under construction in northern California could mean that we're about to find it.

Are we alone?

The modern search for extraterrestrial intelligence (SETI) began 40 years ago, when two US physicists, Giuseppe Cocconi and Philip Morrison, published an article in the science journal *Nature* outlining how microwaves could be used to communicate between the stars. At about the same time, a young US radio astronomer called Frank Drake independently reached the same conclusions. In 1960, he applied his ideas, pointing the 28m (90 ft) Green Bank radio telescope, in West Virginia, at two Sun-like stars to look for any microwaves that could be construed as a signal from ET.

While Drake's so-called Project Ozma found nothing, he did succeed in getting the attention of the rest of the astronomical community. In 1961, he organized the world's first SETI conference, at which scientists from around the world gathered to assess the chances of there really being intelligent life elsewhere in the Universe. In the early 1970s, NASA's Ames Research Center, in Mountain View, California, commissioned an external team of scientists to perform a feasibility study of SETI search methods. Their report, dubbed Project Cyclops, was optimistic about the chances of making contact with extraterrestrial life. Much of the SETI work done today is based on its findings.

By the end of the 1970s, Ames and the NASA Jet Propulsion Laboratory, in Pasadena, California, had established active SETI research programmes. A number of universities had also set up their own projects. In 1988, NASA officials finally approved the plans made by their scientific researchers and granted funds for observational work. Four years later the observations began, but shortly after that the US Congress cancelled the project in a round of budget cuts that also saw the Superconducting Super Collider (SSC) – a giant particle accelerator, that would have been the most powerful in the world – coming under the axe. While the SSC was due to cost a hefty $5 billion, SETI took up just 0.1 per cent of NASA's total budget, just 5 cents per American tax payer per year. The late space visionary Arthur C. Clarke cited the move as firm evidence that there was no intelligent life in Washington DC. But rather than rolling over and dying, SETI has been kept alive by a number of independent organizations. The SETI League is pulling together radio astronomers the world over to monitor the entire sky for signals from ET, while the SETI Institute is using some of the world's biggest radio telescopes to scrutinize specially targeted Sun-like stars.

Where to look?

SETI astronomers are looking for radio signals that fall in the range of frequencies between 1,000 and 3,000 megahertz (MHz, or million oscillations per second), referred to collectively as microwaves. They are the same sorts of waves that bounce around inside your microwave oven. While stars emit lots of visible light and other sorts of radiation, the galaxy is relatively quiet at microwave frequencies, making it a sensible frequency for ET to transmit at. Quiet, that is, with one exception. Microwaves at a frequency 1,420 MHz are emitted in copious amounts by clouds of hydrogen in space. Researchers believe that this frequency will serve as a cosmic bookmark, and are searching the nearby, quieter microwave frequencies.

What form will the first contact with an extraterrestrial civilization take? Current surveys are just looking for signals that span a very narrow range of frequencies that are too narrow to have been produced by any natural phenomena. Astronomers know that any signal they detect that is narrower than about 300 Hz must be artificial in origin, because nature simply can't generate frequencies that precisely.

How many ETIs?

SETI researchers are confident that they're not searching in vain, and that there is intelligent life out there somewhere. There are roughly 400 billion stars in our galaxy. Recent astronomical observations have found planets around a large number of these, boosting scientists' suspicions that planet formation is a common process throughout the Universe. What are the odds of life forming on these planets? And if life is there, what about intelligence? When Frank Drake was planning the first SETI conference, he set about gauging how likely it is that there are ETIs in our galaxy. The result was the Drake equation, a mathematical formula that uses quantities from cosmology, biology, technology and sociology to predict the number of extraterrestrial civilizations living in the Milky Way. Drake calculated that there could be millions of intelligent civilizations in our galaxy. Others have taken a more sceptical view. The Italian–US physicist Enrico Fermi famously asked the question, in 1950, 'Where are they?' His argument was that given the age of the Universe, if there are intelligent civilizations living within it then they should have arrived at our Solar System by now. The fact that we don't see them, argued Fermi, is evidence that they don't exist. The late US astronomer Carl Sagan countered by pointing out that, 'Absence of evidence is not evidence of absence.'

From the ashes

After Congress withdrew funding for NASA's SETI research, Frank Drake set up the SETI Institute, a privately funded organization, to continue the search, based in Mountain View, California. The principal activity of the SETI Institute at this time was Project Phoenix – a targeted search of 1,000 Sun-like stars, all within 200 light years of Earth. The project worked by dividing the microwave band into two billion 1 Hz channels and searching each one in turn for an unusually strong signal. Observations began in 1995, using the 70 m (230 ft) Parkes radio telescope in New South

Wales, then transferred to the giant 300 m (1,000 ft) Arecibo radio dish in Puerto Rico. In 1998, the UK's Lovell telescope at Jodrell Bank, the third-largest steerable radio telescope in the world, joined the project. Lovell's role was to determine whether potential 'hits' detected by Arecibo were manmade or extraterrestrial. Project Phoenix ended in 2004 after failing to find any interesting signals in our part of the galaxy.

The more powerful a telescope is, the smaller the area of the sky that it can monitor. Project Phoenix used the most powerful radio telescopes in the world and so could only monitor very small portions of sky. The institute is now working on a new project, the Allen Telescope Array (ATA), after Microsoft co-founder Paul Allen, who stumped up $25 million to get the project off the ground. It is an interlinked network of 350 6 m (20 ft) radio dishes – located at Hat Creek Observatory, 450 km (300 miles) north-east of San Francisco, California. These dishes will combine to give the power of a single radio telescope 100 m (330 ft) across that's also fully steerable, allowing it to sweep large portions of the sky. By contrast, much of the observational work for Project Phoenix was carried out at Arecibo, which is a fixed-dish telescope. The ATA will be able to listen for alien signals coming from five times further away than Project Phoenix was able to detect, a range of nearly 1,000 light years.

The Allen Array works on a principle called interferometry, where the signals gathered by two or more radio telescopes positioned a distance 'D' apart can be combined by a computer to form an image as detailed as one you might expect from a single telescope dish of diameter D. The ATA is already up and running with 42 dishes operational and gathering data.

DIY SETI

Not all efforts to hunt down ET require vast professional-grade telescopes and millions of dollars of funding. Lower-power telescopes naturally have a wider field of view, enabling them to scan huge swathes of sky, albeit at lower sensitivity than the narrow-field, high-power searches conducted by the SETI Institute. And this is the tack adopted by the SETI League.

The League comprises 1,500 amateur and professional radio astronomers from 62 countries. Each participating member has a radio telescope in their back garden and a PC to analyse the results of their observations. Although modest by professional standards, a typical SETI League amateur set-up uses a dish between 3–5 m (10–16 ft) across. SETI League headquarters coordinates the project, designating each observer their own patch of sky to monitor. The programme aims to monitor the whole sky. Although this would require around a million Arecibo-scale telescopes, scientists hope to achieve all-sky coverage with just 5,000 low-power amateur instruments. This has the advantage over a targeted search of being able to spot life signs from planets that we may not be aware of, orbiting distant stars.

The 'Wow!' signal

The closest thing to alien contact that astronomers have seen so far was spotted over three decades ago. Back in 1977, a SETI researcher at Ohio State Radio Observatory wrote 'Wow' next to a huge radio-emission peak on a data print-out. Known from there-on as the 'Wow!' signal, it was never seen again and so remains unverified. It was eventually written off as radio interference. Even earlier, back in 1967, two British researchers who weren't even looking for aliens thought for a brief time that they'd found them. Jocelyn Bell and Anthony Hewish found pulses of radio emission from space that were so regular, flicking on and off every second or so, that they thought the signal had to be produced by some form of intelligence. The truth was a little more prosaic, though still a significant discovery. Instead of aliens, Bell and Hewish had in fact discovered the first pulsar, a rapidly rotating neutron star (see *How to survive falling into a black hole*) which beams its radiation out into space rather like a lighthouse.

Mr President...

There is a set of protocols that must be adhered to if and when alien contact is made. First, the team making the claim must verify that ETI is the most plausible explanation for the source of their signal. Next the findings are submitted for peer review, where other astronomers are given the team's data and asked to verify or refute the claim. If confirmed, the Central Bureau for Astronomical Telegrams of the International Astronomical Union (who will pass the news to astronomers worldwide) and the Secretary General of the United Nations are both notified. The press is briefed shortly after. In the long term, the signal will be constantly monitored and the transmission frequency protected by international law. But will we reply?

The truth is that we already have replied. Television and radio signals have been spreading out into space for the last 60 years. Any alien civilizations within 60 light years of Earth and with sensitive enough detection equipment pointed in our direction may already be aware of our presence. But not all alien communication is accidental. One signal was sent from the Arecibo dish in 1974 towards a globular star cluster called M13. The signal contained information about our Solar System and life on Earth. However, M13 is 25,000 light years away, so we can't expect a reply for at least 50,000 years.

Whether or not astronomers would reply to a genuine SETI detection is a matter for debate. Many scientists – including Professor Stephen Hawking – have voiced their concern about announcing our presence in this way, arguing that we cannot be sure all life in the Universe will be peaceful. Indeed, if Hawking and colleagues are right, extending the hand of friendship to ET could instead be inviting war of the worlds.

CHAPTER 29

How to make energy from nothing

- Endless energy
- Spinning black holes
- Hawking radiation
- The free lunch universe
- Quantum ratchet

History is replete with examples of optimistic inventors trying to get something for nothing – to build a device that can conjure energy from thin air. Scientists insist that such perpetual motion machines are impossible as they go against the fundamental tenets of physics. But one scientist thinks there may be a way around this problem. When it comes to generating energy, there may be such a thing as a free lunch after all.

Endless energy

From ever-turning water wheels in the works of mind-bending artist M.C. Escher to modern claims for devices that can extract never-ending energy from the behaviour of electric currents in magnetic fields, perpetual motion machines have been rolling off inventors' drawing boards since almost the year dot. True perpetual motion machines, which can output more energy than is put into them, are impossible. They violate a fundamental principle of physics known as the conservation of energy. This essentially says that energy can be neither created nor destroyed, merely changed from one form into another. For example, when you apply the brakes in your car, the car's kinetic energy is not destroyed. Instead, it is turned into heat and sound, which is then shed into the air rushing over the car's brake discs as it moves forward.

Conservation of energy is a keystone in every aspect of physics and has been tested by experiment perhaps more than any other physical law. So how is it possible to create energy from nothing? The secret lies in what you define 'nothing' to be. Most of us would say empty space is nothing. But it turns out that empty space is in fact anything but empty.

Spinning black holes

One source of free energy that's done the rounds in the scientific literature is black holes. In 1963, mathematician Roy Kerr solved the equations of general relativity describing the space and time surrounding a rotating black hole. A static black hole would consist of a spherical event horizon surrounding a gravitational singularity – a point where the force of gravity and the curvature of space both become infinite. However, Kerr was able to show that the space around a spinning black hole is significantly different (see *How to survive falling into a black hole*). In particular, he found that as well as an event horizon – from which there is no return – a rotating black hole is also surrounded by an ergosphere. This is a flattened spheroidal area (shaped like a basketball that's been squashed in a vice) within which space gets swept around with the black hole as it spins.

This effect is known as 'frame dragging', and can be imagined as rather like sticking a spoon in a jar of treacle and twisting it – the treacle near the spoon gets dragged around with it. Frame dragging was first derived from Einstein's equations in 1918, when Austrian physicists Hans Thirring and Josef Lense showed that all spinning bodies exert this effect on the space surrounding them. Just outside a spinning black hole, however, the effect is extremely strong. In 1969, British mathematician Roger Penrose worked out that it was possible to extract energy from the ergosphere of a spinning black hole. He devised a scheme whereby an object entering the ergosphere could get spun up by the frame dragging effect and leave with more energy than it came in with.

Penrose's idea was rather like grabbing hold of a spinning merry-go-round and then jumping off with more speed than you originally had. Only in this case it wasn't quite so simple. Anything entering the ergosphere would need to shed some mass while in there, which would then fall into the black hole. Penrose imagined that a futuristic civilization could use the process by using a rotating black hole as somewhere to dump its garbage. Shuttle craft would periodically carry garbage to the hole. Flying into the ergosphere and jettisoning this material would give the shuttle craft a kick, a bit like the way a rifle kicks back against your shoulder as it discharges a bullet. The fast-moving empty shuttles would then need to dock with some kind of giant dynamo structure that could convert their kinetic energy into usable electricity.

In theory it's possible to extract up to 29 per cent of the energy locked away in the black hole in this way. Some physicists even believe the process could go some way to explaining the powerful jets of material that are seen spewing from the cores of some galaxies – most galaxies are already known to harbour enormous spinning black holes at their centres.

Hawking radiation

In the early 1970s, Cambridge University astrophysicist Stephen Hawking showed that even black holes that are not rotating give off energy. Hawking proved mathematically that black holes should emit particles, effectively evaporating like a body at a temperature determined by the hole's mass. This seems to go against the idea that nothing which has crossed a black hole's event horizon can ever return to see the outside world again. However, Hawking used the rules of quantum theory – the physics of subatomic particles – to arrive at his result. General relativity makes no provision for quantum laws, so it shouldn't really be much of a

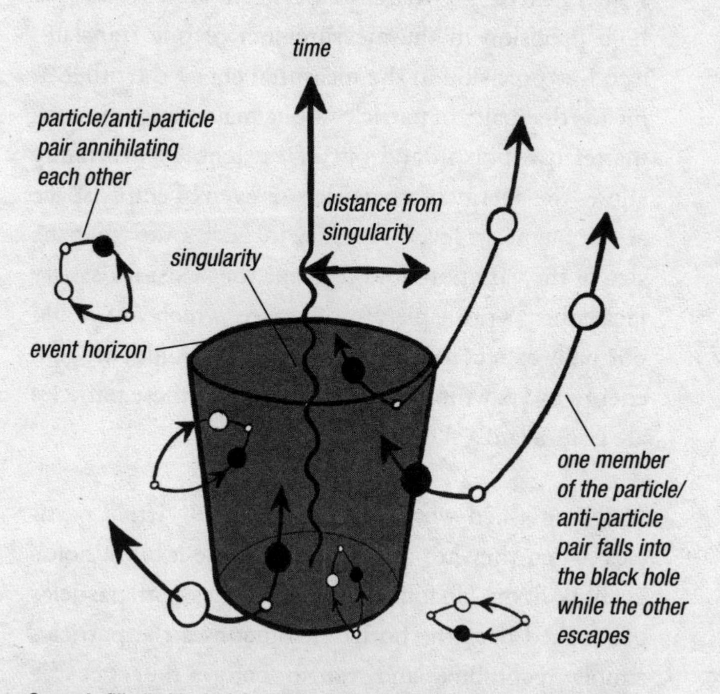

Space is filled with virtual particle pairs popping in and out of existence. But if one partner of the pair falls into the black hole before they can recombine while the other escapes, the net effect is a steady stream of particles from the hole – called Hawking radiation.

surprise to find that quantum considerations can occasionally overturn its predictions.

Hawking's mechanism worked using an idea known as the uncertainty principle. Put forward by German physicist Werner Heisenberg in 1927, it basically says that it's impossible to know both the energy of a particle and the time at which you measure that energy. Instead there is a trade-off between the two so that high precision in the measurement of one translates into low precision in the measurement of the other. It means that pairs of particles – one matter and one antimatter can pop in and out of existence. Uncertainty allows the energy of a particle – or even of empty space at the quantum level – to vary in such a way that the size of the variations and the time they exist for satisfy Heisenberg's principle. Empty space, which you would normally expect to have zero energy, can suddenly gain energy by spawning particles, so long as these particles are gone again a short time later.

Hawking asked what happens to these 'virtual particles' when they are created just outside a black hole's event horizon. He found that sometimes both particles get sucked over the horizon; sometimes the particles simply recombine and vanish before they get the chance to do anything; but sometimes one particle will fall over the horizon while the other has just enough energy to get away from the black hole's gravity. And this creates a steady flow of particles away from the black hole. Meanwhile, the hole's mass – in the absence of any other matter falling in – steadily diminishes. US physicists Lois Crane and Shawn Westmoreland have even suggested that this energy output could be harnessed to run a spacecraft on tiny black holes that have been artificially generated using giant lasers.

Although energy that's extracted from around a black hole will have been mined from 'empty space', some people might argue that it's not strictly energy from nothing – because you need to have a black hole there in the first place. If that's your view then cosmologists have news for you. They say that our entire Universe may have been created from nothing – in the most literal sense of the word possible.

The free lunch universe

It was the day Einstein nearly got run over. He and his colleague the Russian–US physicist George Gamow were out walking in Princeton, New Jersey, one afternoon during the early 1940s. Gamow was explaining to Einstein how one of his students had just calculated that it's possible to make a star from nothing because its mass energy (as calculated from Einstein's formula $E=mc^2$) is exactly equal but opposite to its 'gravitational potential energy'.

The gravitational potential energy of a star or planet is the energy required to assemble it if all its constituent parts were scattered an infinite distance apart. Another way to think of it is as the opposite of the energy that must be delivered to the star or planet in order to completely blow it apart. That energy is positive. And therefore the gravitational potential energy, which is equal yet opposite, must be negative. When Gamow told Einstein about his student's calculation the father of relativity is said to have stopped dead in his tracks. He and Gamow were crossing a road at the time and several cars had to swerve to avoid them. Einstein had realized that the calculation didn't just apply to stars but to the Universe at large as well.

Most cosmologists now believe this is how our Universe popped into creation, probably as a quantum event, much like the virtual particles flitting in and out of existence in empty space that give rise to Hawking radiation. Amazingly though, it wasn't just the matter that makes up the stars, planets and galaxies of our Universe that were created in the 'Big Bang', as the birth of the Universe has come to be known, but the very fabric of space itself. Before that – well, there was no 'before that' because time was created in the Big Bang too, and asking what happened before time was created is like asking what's north of the north pole.

Quantum ratchet

Of course, creating a Universe isn't something that would be remotely useful to you or me, nor is it something we're ever likely to be able to take a crack at. But a physicist in Germany has come up with a way to extract useful energy from empty space itself. It all comes down to the virtual particles we met back in the discussion about Hawking radiation. Back in the 1940s, a Dutch scientist called Hendrik Casimir worked out that these virtual particles would cause two metal plates placed a very short distance apart in a vacuum to move together. It's called the Casimir effect (see *How to travel through time*) and was verified experimentally in 1997. Dr Thorsten Emig has found a way to extract useful energy from the effect via a ratchet-like device that uses the Casimir force to generate rotational motion in one direction, which can then be harnessed. Emig's design works by substituting the smooth plates of the standard Casimir experiment for corrugated ones, which introduces a lateral force that makes the plates slip past one another. By making the corrugations asymmetric, Emig keeps this slipping motion in one direction.

Emig believes the ratchet could generate enough energy to power tiny nanorobots – machines measuring just one ten-thousandth of a millimetre across, which have a host of applications in medicine and for engineering on the smallest scales. The work isn't just another theoretical pipe dream: a real lateral Casimir force has already been measured by a team of experimental physicists working at the University of California. The energy for the virtual particles on which the Casimir effect relies is effectively borrowed from the vacuum of empty space – so Dr Emig's device is, quite literally, mining useful energy from nothing.

CHAPTER 30

How to generate a force field

- Field theory
- The forces of nature
- Quantum fields forever
- Raise shields!
- Electric armour
- Magnetic deflectors

No science fiction spacecraft would be complete without deflector shields to stave off attacks from hostile aliens. Now this idea is gaining traction in science fact. Physicists have come up with designs for force fields to guard real spacecraft against the harsh radiation in space that could otherwise hamper manned missions to planets such as Mars. Meanwhile, other scientists are developing electric shields for tanks and other military vehicles on Earth.

Field theory

Fields are not solely the province of science fiction. Any kind of phenomenon in nature that's able to exert 'action at a distance' does so through a field – a distribution of mass or energy around an object that acts as the source. Perhaps the most familiar example is gravity. Any massive object creates a gravitational field around it that influences other massive objects passing through the field. This is why we have planets in orbit around the Sun – and why cricket balls thrown up into the air come back down to Earth again.

The behaviour of a field in physics is given by an ominous-sounding mathematical entity, known as the field equation. Every theory has its own field equation (and, indeed, sometimes more than one) describing how the field works. The first theory of the gravitational field was put forward in the late 17th century by the English polymath Isaac Newton. The field equation in this theory was a relatively simple law, which said that the gravitational field of a body increases in proportion to the body's mass (so heavy things exert more gravity) and decreases with the distance squared from the source – so if you move twice as far away from an object, its gravitational field diminishes by a factor of four. Newtonian gravity was replaced in 1915 by a far more complex theory – the general theory of relativity – put forward by Albert Einstein. The theory ascribed gravity to bending of space. This in itself was a step up from Newtonian gravity because it explained how the field was being transmitted through space. The field equations of Einstein's theory (there are 10 of them) assert that gravitational interactions propagate outwards as ripples in space and time, travelling at the speed of light.

The forces of nature

Gravity isn't the only player in physics. As far as we understand it, nature is home to a total of four fundamental forces. Gravity, as we've seen, is one. It's mediated by a long-range field that extends right across the Universe, causing far-flung galaxies to move under each other's influence. But despite this, gravity is really quite a feeble force. After all, it takes the entire mass of Earth to keep us humans stuck to the planet's surface. A much stronger force is electromagnetism, the theory of which was developed in the 19th century by British physicist James Clerk Maxwell. The electromagnetic force arises from the interaction between electric and magnetic fields. Maxwell's field equations are four mathematical expressions that describe how moving electric charges generate these fields and interact through them. According to Maxwell's theory, the fields themselves comprise electromagnetic waves – the same entities that light, X-rays and radio signals are made up of. Electromagnetism is much more potent than gravity, exerting a force that's 100 billion billion billion billion (a 1 followed by 38 zeroes) times stronger.

But change was in the air for Maxwell's theory too. In the early 20th century, physicists developing the new discipline of quantum mechanics realized that electromagnetic waves aren't purely waves but can also be regarded to some extent as particles, which they called photons. In 1927, British physicist Paul Dirac worked out a quantum mechanical equation describing the dynamics of negatively charged electrons in the presence of an electromagnetic field. Dirac had taken the first steps towards constructing a theory of electromagnetism that was consistent with the quantum revolution. In the 1940s, the theory was refined and developed to yield a fully 'quantized' version of Maxwell's electromagnetic theory, which was able to describe the behaviour of electrically charged subatomic particles in electric and magnetic fields. Called quantum electrodynamics, or QED, it was the first example of a quantum field theory.

Quantum fields forever

Quantum field theory is at the root of the final two forces of nature. These forces only operate within the nuclei of atoms and are thus inherently quantum – they do not exist in a non-quantized form. The first is known as the weak nuclear force. This is the force that's responsible for so-called 'beta decay' of atoms – a kind of radioactive decay where an atomic nucleus converts some of its particles into either electrons or the electron's antiparticle, the positron. The weak force is mediated by a field made up of two particles – named the *W* and *Z*. The *W* is electrically charged, like the electron, but can be either positive or negative; the *Z* is electrically neutral. The weak force is 100 billion times weaker than electromagnetism. In the late 1960s, a unified description encapsulating both the weak force and QED was put forward and confirmed by experiments. It is still missing one crucial component, the Higgs boson, which particle accelerators such as the Large Hadron Collider are currently hunting for.

We've had the weak force, so it's probably no surprise that the final force of nature is known as the strong force. It too operates exclusively within the nuclei of atoms, and it's responsible for binding the protons and neutrons inside atomic nuclei together. But it soon emerged that the true picture was even more complex. Experiments in the 1960s revealed that the particles in the nucleus are not fundamental entities but are instead made of smaller particles, known as quarks. Each proton and neutron in an atomic nucleus is in fact a cluster of three of these quarks, bound together by the strong force. Whereas electrons and protons carry electrical charge, each quark carries a so-called 'colour charge'. Colour comes in three types – red, green and blue – and each can be either positive or negative. Quantum colour has nothing to do with colour in the real world. It is just a name given to a quantity that is hard to think about – unlike electric charge, colour charge exists purely at the quantum level and we thus have no intuitive grasp of it.

Colour charge generates a field through which the strong nuclear force is mediated. Whereas the electromagnetic field is carried by photons, the colour field is carried by a new kind of particle known as a gluon – of which there are eight different sub-varieties. The theory of colour charge is known as quantum chromodynamics, or QCD. It was developed in the 1960s and '70s and was confirmed by experiments in particle accelerators soon after. As its name suggests, the strong force is the most powerful of all the forces of nature, coming in at about 100 times stronger than electromagnetism.

Raise shields!

Could any of these forces be used to build a working deflector shield? In science fiction, writers often invoke hand-wavy arguments by which a force field could work based on gravity. The basic idea is similar to a gravitational lens (see *How to see the other side of the Universe*), where space is curved by the gravity of a massive object to deflect a light beam (such as a laser blast), or any other inbound object for that matter. Gravitational lenses curve light inwards to focus it. A deflector shield would need to do the opposite and deflect the light away from the spacecraft. One way this could work is using a kind of anti-gravitating material, similar to the dark energy that is thought to pervade the Universe, making its rate of expansion get progressively quicker. According to general relativity, dark energy would curve the space around the spacecraft and deflect incoming objects.

The problem with this idea is that no one knows how to capture and bottle dark energy. Something similar has been produced in laboratory experiments to investigate the Casimir effect (see *How to travel through time*), although only in minuscule amounts. And for the material to bend space enough to be useful, planet-sized amounts of the stuff would be needed. And that's the real problem – gravity is just too weak. What about the weak and strong nuclear forces? These are a no-go right from the off because their range is limited to within the atomic nucleus, which is about a trillonth of a millimetre across. That just leaves electromagnetism – which, it turns out, could be just the ticket.

Electric armour

Scientists at the UK's Defence Science and Technology (DSTL) Laboratory have figured out how to use electric fields to protect tanks from rocket-propelled grenade (RPG) attacks. Fired from a shoulder-mounted launcher, each of these missiles packs a shaped high-explosive charge that melts a mass of copper and then injects the molten material as a jet into the target. Capable of punching through 30 cm (12 in) of steel, these weapons cost just a few hundred dollars and pose a lethal threat to tank crews. The DSTL believe high-strength electric fields could offer a solution. Their design uses a device called a supercapacitor. Capacitors are electrical components that can store up charge and then release it in a massive burst. They are used, for example, in camera flashes. Supercapacitors are a new design that takes advantage of nanotechnology – engineering on tiny length scales of just a billionth of a metre – to store thousands of times more charge than has previously been possible. Electric armour works by using a computer monitoring system that can discharge a supercapacitor into the metal body of an armoured vehicle the moment it detects an RPG launch. This momentarily sets up a huge electromagnetic field around the vehicle, deflecting any inbound metal objects away. The team believe the armour could not only make tanks safer, but also lighter and more manoeuvrable as they would be able to dispense with much of their heavy steel plating.

Magnetic deflectors

Electromagnetic shields aren't confined to planet Earth. British physicist Ruth Bamford has worked out how to surround a spacecraft with a magnetic field that can repel high-energy radiation particles. These particles spew from the surface of the Sun and can cause severe radiation sickness and even death in astronauts. This is something of a show-stopper if, for example, over the coming decades we want to send crewed missions to Mars. One solution is to clad spacecraft with layer upon layer of lead shielding to block this radiation. But lead shielding is extremely heavy and so adds significantly to the weight that has to be blasted into space, making the cost prohibitive. Most of the dangerous particles are electrically charged. Ruth Bamford's idea exploits this fact by replacing bulky lead shielding with a magnetic bubble surrounding the spacecraft, which – just as Earth's magnetic field keeps us safe from cosmic radiation – bats these harmful particles away from the spacecraft and back off into space.

This isn't an entirely new idea, but it was previously thought that an enormous magnetic bubble was required – about 20 km (12 miles) across – demanding a huge electromagnet and power-generation equipment just as bulky as the lead shielding that it's meant to replace. Bamford's calculations now indicate that a bubble just 100 m (300 ft) in diameter would do the trick. The machinery needed to create this could practically fit inside an astronaut's hand luggage and, most importantly, can be built with existing technology. Dr Bamford's idea may not be quite up to fending off Klingon invaders just yet. But it's one small step in the right direction.

CHAPTER 31

How to predict the stock market

- The dismal science
- The futures market
- Sharing the wealth
- Quantum games
- Prediction markets

At first sight, physics and economics sound like they have about as much in common as space flight and potatoes. But an increasing number of researchers are discovering that powerful economic insights can be gained by applying the laws and principles of physics to the movement of money. This field of science now has its own name: econophysics.

The dismal science

Economics is the science of trade. It governs the exchange of goods and services between people, businesses and nation states. Specialists in other scientific disciplines sometimes refer to it as the 'dismal science' because they say it has none of the perceived beauty of the 'natural' sciences such as chemistry, biology and, of course, physics. But in the mid-1990s that disdain began to evaporate and physicists started applying their knowledge of the fundamental behaviour of the natural world and the techniques of physics to try to solve problems in economics and finance.

The application of maths to economics goes back much further, as far as the 17th century. Powerful mathematical methods – such as differential calculus, a branch of maths dealing with how quantities change with time – were employed from the end of the 19th century to allow economists to draw up precise models of how economic systems behave in response to particular inputs. A simple example is supply and demand. As the available quantity of a product or service decreases, so people are prepared to pay more for it – this is 'demand'. Likewise, the more a manufacturer or service provider can charge for a product, the more of it they're willing to sell, in order to maximize profit – this is 'supply'. Plot these on a graph of price against quantity and demand is a downward sloping curve, while supply is an upward sloping curve. Where they cross is called the 'equilibrium point' – and it is towards this point that the actual price of particular goods or services will converge.

There are also vastly more complex cases. For example, Elliott wave theory tries to explain the movements of financial markets in terms of waves of optimism and pessimism in the eyes of investors. These swings create ripples in the prices of stocks and shares, which mathematical economists try to predict. Meanwhile, if you're a hedge fund manager, complex mathematical analysis is all in a day's work. Hedge funds trade in a range of commodities – investing in shares that look set to perform well and 'short selling' stock they believe will fall. The analysis enables the fund manager to literally 'hedge his bets' – spreading the investment in such a way that no matter how the market performs the fund continues to grow.

The futures market

One such market commodity that hedge funds deal with is known as 'futures'. Here, traders don't buy actual shares in a company but rather they buy the option to buy shares for a fixed price at a certain date in the future. If at that time the actual value of the shares is more than the fixed price then the trader can exercise the option and sell the shares straight away for an immediate profit. If, on the other hand, the price of the actual stock is less then the trader can decline to buy, but loses whatever they paid for the option. The behaviour of the futures market is notoriously hard to predict. Traders attempt to rein in this variance by using a formula known as the Black–Scholes equation. This is a fiendishly complicated mathematical relationship linking the price of a stock and the price of the option to buy that stock, taking into account other economic parameters such as interest rates and the market volatility. Solutions to the equation reveal what the maximum price of an option should be for a buyer to make a profit, and what the minimum price is that a seller should offer.

In 1996, physicist Kirill Ilinski at the University of Birmingham in the UK invoked quantum physics to improve on this revered formula. He used the mathematics for the theory of quantum electrodynamics (QED) – the quantum model of electromagnetic fields, which describes the behaviour of electrically charged subatomic particles. Rather than using the formulae to compute the behaviour of positive and negative electrical charges, he switched these quantities for positive and negative amounts of money: credit and debt. He swapped the electromagnetic field – which in QED mediates the interaction between positive and negative charges – for a so-called 'field of arbitrage', which contains all of the information about pricing and interest rates, and thus mediates how credit and debt interact.

Quantum theories like QED obey the uncertainty principle. This means they cannot predict the exact outcome of an experiment, only the probability of each possible outcome. Ilinski used this to model the unpredictability of the stock exchange. And it seems to work.

The Black–Scholes equation emerged from his QED-based formalism, but only after some simplifications. In QED, empty space isn't really empty. Instead, it is full of 'virtual particles' of the electromagnetic field, which randomly pop in and out of existence in accordance with quantum uncertainty. In the financial version of the theory, virtual particles of the arbitrage field also exist and play the role of random opportunities. But in the same way that other quantum effects conspire to quickly damp out virtual particles in QED, virtual opportunities in arbitrage are also extremely short-lived.

Ilinski says this can be interpreted as the presence of speculators in his model. Speculators anticipate price changes and act quickly when an opportunity arises, rapidly erasing the chance for others to benefit from it. In the absence of speculators, Ilinski recovered the standard form of the Black–Scholes equation. But leaving speculators in, via their QED counterparts, gives an extended version of the formula that makes it possible to hedge deals against the actions of these market traders.

Sharing the wealth

It is sometimes said that in any economy 20 per cent of the people own 80 per cent of the wealth. Now physics is shedding light on this too. One of the first people to carry out a mathematical analysis of the distribution of wealth in a population was the French engineer Vilfredo Pareto. In 1897, he figured out that – in Europe at least – money was distributed according to what's known as a power law. In other words, the number of people with more than a given amount of wealth, W, was proportional to $1 / W^e$ – where the 'power' e, Pareto found, is a number varying between 2 and 3. It meant, essentially, that the number of people with a lot of money was very small. Since then, economists have realized that Pareto's law only applies for large values of W, corresponding to the top three per cent of the population. The bank balances of everyone else must follow a different rule.

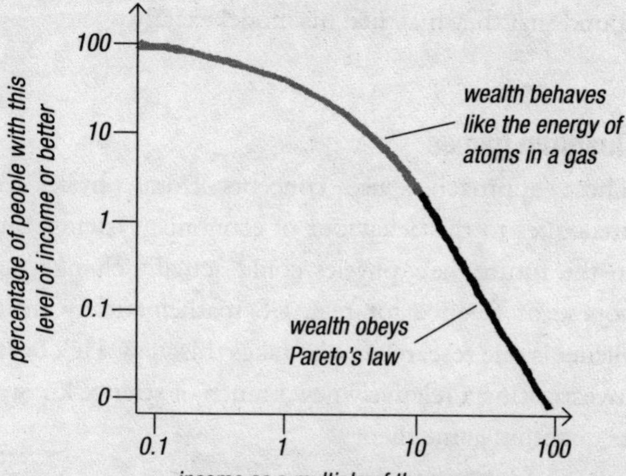

The wealth of the richest 3 percent obeys a power law, known as Pareto's law. For the rest of the population, their income mirrors the behaviour of atoms in a gas.

In 2005, US physicist Victor Yakovenko decided to work out exactly what this rule is. He based his reasoning on a rather unlikely model: the behaviour of atoms in a gas. Yakovenko realized that money is rather like energy, in that it can never be destroyed – but simply flows from one place to another. He thus modelled the financial dealings of a large group of people using the same mathematical formulae used to describe atoms of gas colliding with one another and exchanging energy. The same formula from the gas theory giving the number atoms above a certain amount of energy gives the number of people with more than any given amount of wealth. When Yakovenko dug out data on the income of US citizens he found that they matched his model exactly.

Quantum games

These approaches use concepts from physics as analogues to the behaviour of economic systems. But in the future, new physics could actually change the course of finance for real. US mathematician Steve Bleiler is one researcher who takes this view. He's been investigating a relatively new branch of science known as quantum game theory.

Ordinary game theory, as the name suggests, is the branch of maths that deals with selecting the optimal strategies players should use to maximize their return when playing a game. It works by assigning every possible strategy a numerical value known as the 'pay off'. The best strategy to choose is the one that gives the highest pay off in the worst-case scenario – in other words, when your opponent is also playing optimally. This situation, where both sides play their optimal game, is known as a Nash equilibrium, after the US mathematician John Forbes Nash, who put forward the idea in the 1950s. Game theory is used by political campaigners and military planners, and by economists to help them select investment strategies.

The existing formulation of game theory relies on classical laws of information, where data is stored and processed as binary digits, or bits: a stream of 1s and 0s. But in the coming years, classical information processing looks set to give way to its quantum equivalent, using quantum computers that encode information as 'qubits' (which can be both 1 and 0 at the same time), giving them capabilities that far outstrip any desktop PC of today (see *How to crack unbreakable codes*). This new information architecture will usher in a new incarnation of game theory, known as quantum game theory. Bleiler and many other researchers believe that quantum game theory will throw up radically different optimal strategies from those of the classical theory. Traders who are aware of these strategies when doing business over quantum information channels will have a distinct edge over those who don't.

Prediction markets

If the science of how the world works can shed light on economics, then it seems reasonable to wonder whether economics can return the favour. And it seems that it can, through what have become known as prediction markets. These work rather like a stock exchange, but instead of buying and selling shares in companies, traders deal in the outcome of future events. That might be the result of a political election, for example. In this case, traders can invest in virtual shares in each political party, with each share priced between £0 and £1. As politicians hit the campaign trail, the traders can buy or sell their shares. And as they do, market forces drive the share prices to reflect each party's likelihood of getting elected. For example, if the UK Conservative party is trading at £0.42 per share, that translates into a 42 per cent chance of them winning. Once all the votes in the election have been counted, those holding shares in the winner get £1 per share; everyone else gets nothing.

Since 1988, the University of Iowa has run just such a market on the result of US presidential elections. Called the Iowa Electronic Market (IEM), it's been closer to the actual outcome than the opinion polls 74 per cent of the time – and correctly predicted Barack Obama's 2008 victory. With the Internet as a convenient interface for traders, there are now online markets in everything from the weather to the box office performance of movies. The Hollywood Stock Exchange correctly forecast 32 of the 39 nominations for the 2006 Oscars.

Prediction market Intrade is even running a market on the discovery of the Higgs boson (see *How to recreate the Big Bang*), though from the prices they are offering, the odds of it turning up do not look great. Professor Robin Hanson, an economist at George Mason University, Washington DC, even believes betting on scientific issues could accelerate the pace of research – offering extra incentives to scientists and stimulating public interest in science. Hanson admits his idea still needs some development, and indeed a shift in the law – at present betting on scientific issues is illegal in some countries. But if it gets the go-ahead then, while physics could hold the future for economics, economics may well tell us the future of physics.

CHAPTER 32

How to crack unbreakable codes

- Secret society
- Codebreakers
- Public key encryption
- Quantum computing
- Shor's algorithm
- Quantum solace

Codes are a crucial part of military communications and for conducting secure financial transactions across the world's electronic banking networks. Modern encryption techniques are so complex it would take longer than the age of the Universe to crack them using even the fastest computers. Well, almost. Quantum physics is now making possible computers that can decipher the toughest codes in just minutes.

Secret society

Every time you send your bank details over the Internet you are trusting your hard-earned savings to a system that could soon be about as secure as tissue paper. Financial dealings are encrypted, or translated into code, before they are sent, just in case any eavesdroppers happen to be listening in. Most encryption systems rely on ciphers, which are mathematical procedures that are fiendishly hard to unpick, unless you are the intended recipient with the key that tells you how the procedure can be reversed. A simple code might involve switching the letters of the alphabet in your message for different ones. This is known as a substitution cipher. But ciphers rapidly become more complicated. For example, you could use different substitution keys for every letter – rotated in a strict cycle that only you and your recipient know.

Codebreakers

The first example of coded text is believed to have been made in ancient Egypt in 1900 BCE, when a scribe drafted a passage of text using a non-standard set of hieroglyphs. Some of the first cryptanalysts – codebreakers – lived in 16th-century England. The English had gained a reputation for intercepting the communications of foreign diplomats, so many governments began encrypting their important messages. In response, England founded its first intelligence department, dedicated to deciphering these messages.

Perhaps the most notable triumph of England's (now Britain's) codes and cipher effort happened 350 years later, during World War II. In 1938, with war looming, the Government Code & Cipher School (GC&CS) set up a codebreaking centre to decipher enemy transmissions at Bletchley Park, a Victorian mansion in Buckinghamshire. The centre was assigned the codename Station X. The Bletchley codebreakers were some of the greatest mathematical minds of the 20th century. By the end of the war they had broken the Nazis' most sophisticated encryption system: the Enigma.

Public key encryption

Deciphering the Enigma was facilitated in part by the capture of actual Enigma machines and codebooks. Indeed, the danger that a third party may get hold of the key is the weakness of all codes. Or at least it was prior to a discovery in the early 1970s by a team of researchers working at Britain's Government Communications Headquarters (GCHQ). James Ellis, Clifford Cocks and Malcolm Williamson devised an encryption system in which you can tell the key to whomever you like, confident that only the message's intended recipient will be able to read it. Called 'public-key cryptography', the system works using some smart yet simple maths. The intended recipient of the message picks two very large prime numbers, that is numbers which can only be divided by themselves and the number one. These form the secret key, needed to read the message, and aren't revealed to anybody. Multiplying these two numbers together produces what is called the public key, which the recipient broadcasts to all and sundry. Encrypting a message using the code only requires knowledge of the public key, but to decode that message requires the two factors, which are only known to the recipient. It is easy to multiply two big numbers together, but splitting one large number into its factors is phenomenally hard, so the secret key remains secret.

GCHQ's secrecy policy means this system is often credited to a team from MIT, who discovered it several years later. Today, it's best known as RSA, after the three initials of the MIT group's surnames, Rivest, Shamir and Adleman. RSA is one of the most secure encryption systems in use today and is employed widely in electronic commerce. It is very secure. As of April 2010, the largest number factorized is 232 decimal digits long. It took researchers two years using hundreds of computers to get the answer. It is estimated that to factorize a 2,000-digit number would take longer than the estimated time for the Sun to burn itself out. But now RSA looks set to be undermined by a radical new breed of computer – one that runs in parallel universes.

Quantum computing

Quantum computers were first conceived by the British physicist David Deutsch in 1982. Ordinary, or 'classical', computers work by storing data in the form of binary digits, or 'bits', represented by a stream of 1s and 0s. These are stored inside the computer's circuits using electronic switches called transistors – switching the transistor to the 'off' position corresponds to 0, while switching it 'on' records a 1. The data is then manipulated using electronic logic gates, which operate according to the laws of classical electromagnetism. However, on the smallest scales, matter doesn't obey classical electromagnetism. It is dominated instead by the laws of quantum theory. Deutsch realized that if you could build logic gates which were small enough that they operated according to quantum laws, they could permit a new kind of computing machine with unusual and powerful capabilities.

Quantum computation is based not on bits but on qubits, short for 'quantum bit'. Whereas a classical bit can take the value either 0 or 1, a qubit is both 0 and 1 at the same time. This is thanks to a weird property of quantum systems that enables them to exist in all possible states at the same time until the system is actually measured. A number of qubits together make a 'qubyte'. A classical byte, made from 8 bits, can store any one number between 0 and 255. An 8-qubit qubyte, however, stores every number between 0 and 255 at the same time.

Passing a qubyte through a quantum processor allows all 256 different numbers stored in the qubyte to be processed in one go instead of one-by-one. Deutsch calls this 'quantum parallelism'. If he is right, quantum computers run their parallel processes in parallel universes. The technical name for this is the 'many worlds interpretation' of quantum theory. It says that there is an infinite number of universes in which every conceivable eventuality happens (see *How to live forever*). Deutsch has calculated that a quantum processor could easily be 10^{500} times more effective than its classical counterpart – that's a staggeringly large factor of 1 followed by 500 zeroes. But all the data for these parallel calculations must be stored somewhere. The number of atoms in the Universe is only 10^{80}. As each atom can at most store one bit of useful information, there simply aren't enough atoms in our Universe alone to do the calculation. And that's where parallel worlds come in. Deutsch says the computer draws upon the computing power and data storage of its counterparts in these other universes to get tasks done super quick.

Shor's algorithm

The discovery of quantum computers alone wasn't enough to make cryptographers sweat. Just like ordinary PCs, quantum processors need to be programmed. That means coming up with an algorithm – a sequence of steps that the computer can follow in order to get the calculation done. Since the rules of quantum programming were so new, no one quite knew where to start.

That changed in 1994 when a computer scientist called Peter Shor, based in New Jersey, worked out the algorithm that would allow a quantum computer to split a number of any size into its two prime factors. Whereas this calculation can take conventional computers tens of billions of years, a quantum computer may be able to do it in just a few minutes. This means that anyone intercepting a public encryption key could now have the means not only to send messages using that key but also to eavesdrop on the conversations of others.

It all looks great on paper, but actually building a quantum computer is no easy matter. Very basic designs have been constructed in labs. These work by storing data on the quantum states of subatomic particles, such as the nuclei of atoms. Each atomic nucleus has a property called quantum spin, with just two values: 'spin up' and 'spin down'. Crucially, a particle can exist in a superposition of both spin states at the same time, and this makes it ideal for storing the value of a qubit. The system is manipulated by bombarding the atoms with radio pulses. The final states of the qubits are then read out by measuring electromagnetic signals from the nuclei, called resonances. The problem that's holding the researchers up is a phenomenon known as decoherence. This is where a quantum system interacts with its surrounding environment, destroying the delicate balance of purely quantum interactions on which a quantum computer relies.

However, researchers are making progress. In 2009, scientists in Bristol in the UK built a rudimentary quantum processor from silicon. Whereas the silicon chips in your computer work electronically by manipulating electric charges, the Bristol chip is an optical device that acts on individual photons of light. The device was not only free from the effects of decoherence, but the team were able to run Shor's algorithm on it. It may still be some time until we have quantum processors buzzing under our desks, but most researchers believe it will happen one day in the coming decades. When we do, RSA and all other forms of public-key encryption will be rendered obsolete.

Quantum solace

Quantum theory may yet give the cryptographers the last laugh. Today, electronic communications are transmitted as pulses of electric current down a wire, but what if the quantum particles used to store data inside a quantum computer were also used to transmit it? The same sensitivity of quantum systems that gives rise to the phenomenon of decoherence means that an eavesdropper trying to listen in on a message transmitted as a beam of quantum particles must inevitably interfere with the message, altering some of its content. A transmitter and receiver could therefore check that no one has tried to intercept their communications by interspersing the actual message with a test signal – a known sequence of bits. This could be used to distribute an RSA key to intended recipients only, rather than broadcasting it publicly. Any discrepancy in this test signal would mean that the key had been intercepted, in which case the transmitter simply sends a new one. Only once the receiver is in possession of a key that both parties are sure has not been intercepted will the transmitter use it to encrypt and send their actual message. This technology is already here. In 2007, Swiss company id Quantique used just such a protocol to transmit electronic ballot slips during the Geneva federal elections. It remains to be seen whether this will be the last word on cryptography or just the next escalation in the ongoing arms race between those who make codes and those who break them.

CHAPTER 33

How to build an antigravity machine

- Down force
- Quantum gravity
- The rocketmen
- The law of lifters
- Superconductivity
- Podkletnov's disc

Antigravity devices have long been staples of science fiction, blocking the gravitational force to enable you to float in the air as if you were in space. A working antigravity machine would mean safe air travel, pollution-free cars, ultra-efficient elevators and cheap, easy access to Earth's orbit. Various devices already use other fields of nature, mainly electromagnetism, to balance the force of gravity, allowing objects to hover in the air. But one maverick researcher thinks he knows how to switch off gravity itself.

Down force

Gravity is the force that keeps us stuck to the surface of planet Earth. It's the weakest of all the four forces of nature, the others being electromagnetism and the strong and weak nuclear forces that operate inside the nuclei of atoms. Even so, with a whole planet's worth of mass to act as the source, there is still a considerable force pulling on the heels of anyone trying to escape its grasp. This fact is foremost in the minds of rocket engineers. Rockets expend a terrific amount of fuel to escape Earth's gravitational pull, making space travel an expensive and dangerous business.

Antigravity would have the power to change all that. With none of the planet's gravitational pull to overcome, the slightest upward shove would send a spacecraft climbing skywards like a helium balloon. However, as far as we know from observations – and from our best theory of gravity, Einstein's general theory of relativity – all positive concentrations of mass and energy generate attractive gravity. Generating repulsive gravity would demand negative mass or energy. This has been made in very tiny amounts in lab experiments but in order to counteract the gravity of a planet, you'd need a planet-sized blob of negative-mass material. And that's not something you find lying around. There is an alternative possibility, that one day we could try to block the gravitational interaction somehow. Just as shielding sensitive electronic equipment in a lead box can guard it against the potentially harmful effects of electromagnetism (see *How to cause a blackout*), it may also be possible to block gravitational fields.

Quantum gravity

Blocking gravity will require a detailed understanding of how the gravitational interaction is propagated. In general relativity, gravity is caused by the bending of space and time – and there is plenty of observational evidence to suggest that this theory works. Nevertheless, ultimately we will need a quantum theory of gravity: a version of general relativity that works down at the scale of subatomic particles. If you believe the Big Bang theory for the origin and evolution of our Universe – and most professional cosmologists today do – then space expanded from a superdense point. At and around this point, space itself was so small it existed in the quantum domain. And that means we need a quantum theory of gravity to explain its behaviour if we are ever to comprehend how our Universe came to be.

Other theories have been quantized. For example, the 'classical' theory of electromagnetism developed in the 1860s by James Clerk Maxwell and others was adapted a little under a century later to make a full-blown quantum theory of electric and magnetic fields. This new quantum theory was able to describe the interactions between charged subatomic particles in terms of the exchange of photons, particles of the electromagnetic field. The gravitational field, on the other hand, is mediated by particles called gravitons, which are yet to be discovered experimentally. Fathoming their behaviour enough to say how they can be blocked – if at all – will require a full mathematical theory of quantum gravity which, thus far, physicists have been unable to produce. Gravity and quantum theory, at least on paper, are a match made in hell. However, new particle physics models such as string theory and M-theory (see *How to visit the tenth dimension*) could show us the way forward.

The rocketmen

If gravity can't be modified or blocked – at least not using the existing laws of physics – can we bring any other branches of physics to bear in order to counteract it? The answer, of course, is yes. Perhaps the simplest example would be vertical/short take-off and landing (VSTOL) jet fighters. The first, and the most famous, of these is the British-built Harrier jump jet, which made its first flight in 1960.

The challenge was to build a jet engine so powerful that its thrust alone could lift the weight of a fighter aircraft vertically upwards, and to do so using no more fuel than a conventional jet. A team at Bristol Aero-Engines achieved this with their Pegasus engine. They used an ingenious water cooling system, which allowed the engine to be pushed way beyond its normal power rating for short periods. Four directable nozzles channelled its thrust downwards for vertical take-off and landing, and could then be swivelled to point backwards and generate the propulsion needed for forward flight. Along the way, this so-called 'thrust vectoring' led to a new tactic in aerial combat, called VIFFing ('vectoring in forward flight'). Here, fighter aircraft use thrust vectoring to brake rapidly, dramatically improving their manoeuvrability. The technique has been perfected by Harrier pilots in the US Marine Corps. A similar system to VSTOL is used in jet packs, which use a directable rocket motor, most commonly running on hydrogen peroxide. Strapped to the back of an intrepid pilot, the jet pack enables them to hover in mid-air. However, the limitations on the amount of fuel that can be carried and the safety concerns – jet packs fly too low for a parachute to be of any use – mean that there are few of them in use today.

The law of lifters

Roaring jet engines and rocket motors somehow don't seem to capture the spirit of what most of us probably think of at the mention of antigravity. These devices shouldn't be noisy (bar maybe a slight hum), they shouldn't rattle your fillings loose as you fly them, and they definitely shouldn't belch out clouds of hot exhaust gas that threaten to cook passers-by whenever you set down in a crowded street. There is, however, another kind of propulsion device that looks, feels and sounds more like the common conception of an antigravity machine. It is called a lifter and it works on the principles of electromagnetism, taking advantage of a phenomenon known as an ion wind.

You'll occasionally see lifter models demonstrated at science fairs. They usually consist of a triangular balsa wood frame with a ribbon of aluminium foil stretched around the outer edge. Balsa wood posts protrude up from the frame, around which is stretched a length of fine-gauge wire, running a few centimetres above the foil. The lifter is then connected to a high-voltage source – typically around 30,000 V – with the positive terminal of the source clipped to the wire and the negative terminal to the foil.

When the current is switched on, the high voltage pulls negatively charged electron particles away from atoms of gas in the air and towards the positive wire. This leaves the atoms themselves positively charged – a process called ionization. The positive atoms are then repelled away from the wire and accelerated towards the negatively charged foil. As they move, they collide with other uncharged atoms, creating a breeze of air that travels downwards. Newton's third law of motion says that for every action there is an equal and opposite reaction (this is the reason rockets fly) – and the reaction to the downdraft of air is an upward force that makes the lifter rise up and hover. Sadly the amount of payload that a lifter can raise is tiny – about a gram per watt of power applied. This means that you have to make them from balsa wood and foil to be as light as possible. It also means that, given the hefty power supply needed, it's unlikely a lifter could ever fly under its own steam. Nevertheless, they are very cool gizmos indeed.

Superconductivity

Other more sophisticated electromagnetic levitation effects have been discovered. One is called the Meissner effect and arises as a result of superconductivity. This is a property of certain materials whereby they assume a state with zero resistance to electrical current – usually when they are cooled close to absolute zero (−273°C/−459°F). A current in a loop of superconducting wire will, in principle, circulate forever. It has long been known that cooling a material down improves its electrical conductivity. Temperature is caused by vibrations at the atomic and molecular level, so a warm lattice of atoms and molecules has thermal motions that make it jiggle about, impeding the flow of electrons through it. In 1911, Dutch physicist Heike Kamerlingh Onnes demonstrated that this resistance can be made to vanish entirely. The theoretical explanation for the effect would follow in the 1950s. In a nutshell, electrons that are cooled sufficiently lock together to make so-called 'Cooper pairs' that are able to slip freely through the lattice.

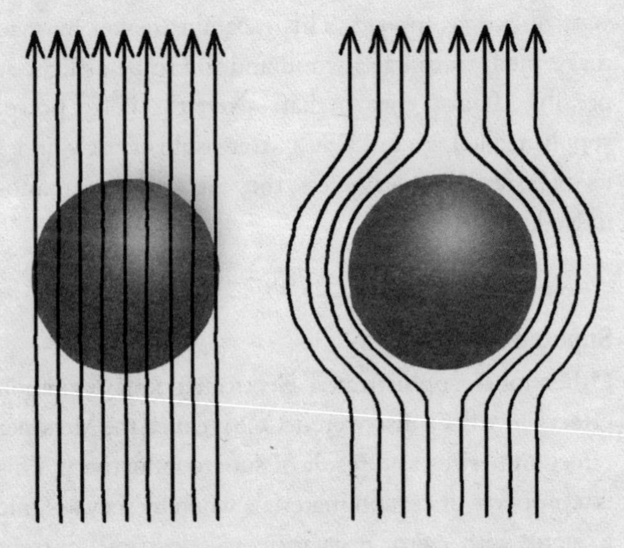

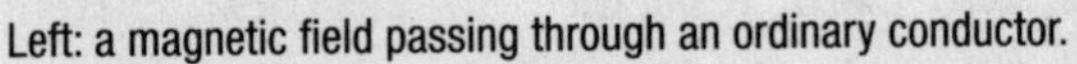

Left: a magnetic field passing through an ordinary conductor.

Right: a magnetic field is expelled from a superconductor by the Meissner effect.

Superconductors are used to build powerful magnets for particle accelerators and medical imaging machines. Physicists are now trying to synthesize materials that can exhibit superconductivity at room temperature. The Meissner effect arises because superconductors expel any magnetic fields that try to pass through them. This happens because the magnetic field induces a current in a conductor by induction (see *How to cause a blackout*). In a superconductor, this current flows over the surface of the material, in turn setting up a magnetic field of its own – again by induction – which exactly cancels out the external field. This means that a magnet placed above a superconductor will levitate, supported in mid-air by an equal and opposite magnetic field.

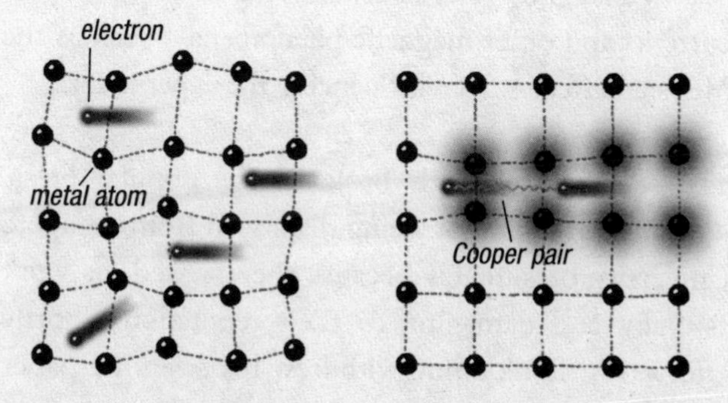

In an ordinary conductor (left), the vibrating lattice of atoms impedes the flow of electrons through it.

In a superconductor (right), the lattice is cooled to minimize vibrations. Meanwhile, electrons bond together to form so-called 'Cooper pairs' which slip through the lattice more easily.

Podkletnov's disc

No round-up of antigravity research would be complete without a mention of the exploits of maverick Russian scientist Eugene Podkletnov. In 1996, Podkletnov, a physicist at Tampere University in Finland, claimed to have built a device that not only counteracts the force of gravity but actually blocks it. Like the Meissner effect, Podkletnov's work made use of superconductors. Specifically, he used a 30 cm (12 in) disc of superconducting yttrium-barium-copper oxide. He said that when this disc was cooled to −230°C (−382°F) and spun up to 5,000 rpm, objects placed above it would lose about 2 per cent of their weight. He was careful to stress that he had gone to considerable lengths to prevent air currents and other magnetic phenomena – such as the Meissner effect – from influencing the experiment.

The story subsequently broke in the popular press, leading to Podkletnov being dismissed from Tampere University, presumably because they viewed his work as flaky and damaging to their reputation. Shortly afterwards, Podkletnov withdrew his scientific paper on the subject from publication. Various groups around the world at universities and private research organizations – and even NASA – have tried to replicate his results, but without success. These attempts were not helped by the fact that Podkletnov was less than forthcoming in giving away the precise details of his experimental set-up.

Nevertheless, in interviews given to the press since, Podkletnov maintains that his device demonstrates a true antigravity effect. He continues to develop the idea with other theoretical and experimental physicists. Some scientists think Podkletnov is dishonest or crazy, or both; others say he's just plain wrong. But if he is right – and there's at least an outside chance that he is – then his findings will change everything from the exploration of deep space to how we travel to work in the morning.

CHAPTER 34

How to create life

- The light elements
- Star formation
- Red giants
- Supernova
- Planet formation
- The anthropic principle
- Space for man

When singer Joni Mitchell stated in her 1970 song 'Woodstock' that 'We are stardust' she may not have been aware that her lyrics were 100 per cent scientifically accurate. Human beings, all other life on Earth, and the very planet itself, are made of material forged in the hearts of stars and then scattered across space in enormous stellar explosions known as supernovae.

The light elements

The Big Bang in which our Universe was created brought into existence space, time and all of the matter that we see around us. But matter can take a variety of forms. The material that our everyday world is made up of – known to scientists as baryonic matter – is divided into various chemical elements. A chemical element is defined by the number of positively charged proton particles in the nucleus that lies at the heart of every one of its atoms. For example, carbon has six protons while oxygen has eight. And these elements can be converted into one another by nuclear reactions (see *How to turn lead into gold*).

Nuclear reactions come in two types: fission, which involves chopping up large atoms; and fusion, which involves joining together smaller ones. In the Big Bang there were no heavy elements, only primitive subatomic particles, so the only reaction possible was fusion. But fusion requires high temperatures in order to slam protons together hard enough to overcome the mutual repulsion they feel as a result of their positive electrical charge. In the Big Bang fireball the temperature was only hot enough to do this for the first few hundred seconds. And this only gave enough time for the formation of the two lightest chemical elements, hydrogen and helium. The Universe after the Big Bang was composed of about 75 per cent hydrogen and 25 per cent helium, with a smattering of heavier elements. Studies of primordial gas clouds in deep space have confirmed these proportions. Life requires much more complicated elements than just hydrogen and helium, including oxygen and carbon, but also nitrogen, calcium, chlorine, potassium, iron and many others. These have to be cooked up in another kind of cosmic nuclear furnace – the cores of stars.

Star formation

The first stars are thought to have formed from collapsing cosmic clouds about 200 million years after the Big Bang. Stars are formed from giant clouds of hydrogen gas drifting in space. These clouds are very rarefied, with densities of no more than a hundred atoms of gas per cubic metre, but they can span hundreds of light years in size, and contain hundreds of thousands of solar masses of material. Tiny irregularities in the density of such a cloud lead to instabilities as over-dense regions begin to collapse under their own gravity. As they collapse and become denser, their gravity increases, causing them to pull in more mass still, increasing their density further in a positive feedback loop.

This ball of gas is known as a protostar. As it collapses, squashing itself, the temperature inside starts to rise. When the core has reached about 15 million °C, nuclear fusion reactions can ignite, combining nuclei of hydrogen to make helium. Energy floods out and a new star is born. The energy heats the gas in the star's outer layers, raising its pressure until the outward force is sufficient to halt the star's inward collapse under gravity. Astrophysicists call this balance point hydrostatic equilibrium. Once a star is in hydrostatic equilibrium, it enters the phase that will take up the bulk of its life. How long this lasts depends on the star's mass. For a star like our own Sun, it will be about 10 billion years. Heavier stars, however, live fast and die young. A star 10 times the mass of the Sun will shine for only 10 million years. The maximum size permissible for any star is about 100 solar masses, at which point the copious radiation streaming out from the star blows it apart.

Red giants

Once a star has run out of hydrogen fusion fuel in its core, a change begins. The core begins to contract. The contraction heats the core to the point where it's hot enough to ignite fusion of helium – which the star has just spent millions or billions of years producing a great deal of – fusing three helium nuclei together to make a nucleus of carbon. The extra energy produced by helium fusion inflates the star's outer surface to 10 times its previous diameter, forming a red giant. Meanwhile the star continues to burn hydrogen in a thin shell around the outside of the core. Sooner or later the supply of helium runs out too and the process repeats. Now carbon ignites in the core and, via a range of fusion reactions, forms elements such as neon, sodium, magnesium and oxygen. Meanwhile helium burning continues in a shell surrounding the core, with hydrogen burning in a shell outside that. This process continues until the core is made of iron, beyond which the binding energy per particle in the nucleus decreases, making nuclear fusion inefficient. At this point, there is nothing to act as nuclear fuel in the core and it begins to cool. The cooling lowers the pressure. Now the weight of the star pushing down makes it shrink, and it implodes.

Supernova

What happens next depends on how heavy the star is. Stars up to about 10 times the mass of our own Sun undergo cycles of oscillation as fuel material from the outer shell sinks down into the core. For example, helium produced by the hydrogen burning shell sinks down until the temperature is high enough for it to ignite. This puffs the star up again until it runs out of fuel once more and falls back into collapse. These oscillations become more violent as the star uses up the last of its fuel. The process culminates in one final outward sigh that flings off the star's envelope to form a diffuse and ghostly cloud of gas called a planetary nebula. At the centre of the nebula is a bright pinprick of light: a dense white dwarf star.

For stars larger than 10 solar masses, the process is not as serene. Such a star will grow to immense proportions during its death throes, becoming what is known as a red supergiant. The collapse of a supergiant's core is unstoppable – until, that is, the core density becomes so great that electrons and positrons merge together to form a super-dense blob of neutrons, known as a neutron star. The quantum mechanical pressure inside the neutron star suddenly halts the collapse, causing a 'bounce' that catapults the star's outer layers off into space in a cataclysmic explosion known as a supernova, with the neutron star left behind as a remnant at its heart.

Planet formation

Both supernovae and the formation of planetary nebulae serve to scatter the chemical elements cooked up inside a star out across space. Subsequent generations of stars can then form from these element-enriched clouds. When these stars die and scatter their material into space the resulting clouds are enriched further. Once the interstellar medium had acquired enough heavy elements, new stars could also develop a belt of dust and debris circling their equators, known as a protoplanetary disc. Gradually the particles of dust stick together to form small rocks. Once big enough, these chunks of material develop gravitational fields strong enough to attract more material, creating mountain-sized chunks and eventually planets. Close to the raging heat of a newly ignited star, rocky planets form as all the gases and other volatile materials get blasted away by the intense radiation. Further away, the temperature is much lower. Here, the volatile materials condense to form gas giants, such as Jupiter and Saturn.

The anthropic principle

We know that on at least one planet in our Solar System a chemical process that we call 'life' has emerged. Life is based on a wide range of chemical elements, all of which were forged in the hot cores of long-dead stars. The proliferation of life on Earth has actually helped in the understanding of how these elements came to be. In the 1950s, the British astrophysicist Fred Hoyle pointed out that in order for there to be enough carbon in the Universe for carbon-based life (such as that found on Earth) to exist, there must be a mechanism within physics for carbon to be mass-produced inside stars.

Making carbon is tricky. It has a total of 6 protons and 6 neutrons in its nucleus. Helium – the material inside stars that the carbon has to be made from – has 2 protons and 2 neutrons. So three helium nuclei need to combine to make each nucleus of carbon – and it is unlikely for three particles to come together at a single point in space. However, two helium nuclei can combine to make the element beryllium – with 4 protons and 4 neutrons. The beryllium can then join with a third helium nucleus, but this shouldn't form carbon because it's unstable – it's got too much energy and breaks apart in a few nanoseconds. Hoyle predicted that there must exist a so-called 'resonance' of carbon, an energized state of the carbon nucleus. It would then be energetically more favourable for the beryllium-helium cluster to turn into this energized state of carbon and then drop back down to become an ordinary carbon nucleus, than for the cluster to break apart. And, sure enough, when experimental physicists went away and looked for Hoyle's resonance, they duly found it. It was cited as an example of what's become known as the 'anthropic principle' – the idea that the laws of physics must be such as to allow carbon-based life-forms to emerge in the Universe because otherwise we would not be here!

Space for man

Some physicists regard this as a powerful principle. Others resent the fact that it's even been elevated to principle status, preferring to use the term anthropic reasoning. One of the leading objections to the anthropic principle is that it suggests, in a slightly religious way, that 'someone' has somehow fine-tuned the laws of physics in our Universe for our benefit. Some physicists believe the so-called many worlds interpretation of quantum theory (see *How to live forever*) could offer a way out of this predicament. Many worlds insists that ours is not the only universe that exists, but just one in a sprawling network of parallel universes that physicists have dubbed the multiverse. In this multiverse view, universes with all possible combinations and permutations of the laws of physics must exist. This would mean that the specific set of laws that prevail in this universe – our Universe – are effectively determined at random.

Once you look at it this way, it's really no surprise that we are here to observe physics to be the way it is. Rather than just one universe in which there's just one chance for the conditions to be right for life, there is a virtually infinite number. Inevitably, in some of these the conditions will, by chance, facilitate life – with no need for God-like fine-tuning. And only in these universes – of which ours is clearly one – will there be physicists scratching their heads and wondering why.

CHAPTER 35

How to read someone's mind

- What is MRI?
- Going functional
- Tell me the truth
- Dream watchers
- Mind control

No longer the preserve of self-proclaimed psychics and frauds, real mind reading has been made possible by physics. It is done using a medical scanning technique known as functional magnetic resonance imaging (fMRI), which was originally designed to diagnose and monitor the growth of tumours and other disorders in the brain. Now fMRI has found a new niche revealing exactly what people are thinking.

What is MRI?

The first ever magnetic resonance image (MRI) was taken in 1973 and the first studies of its use on humans took place in 1977. MRI is preferred over usual X-ray imaging because it's less harmful (X-rays are a very high-energy form of radiation), making it especially suited to the treatment of patients with chronic illnesses, such as cancer, who require many scans. MRI works by measuring radio waves given off by water molecules in the body. It does this by using a magnetic field to stimulate the protons that lie at the centre of each hydrogen atom inside the water molecules. First, a huge magnetic field is applied to the body. It causes positively charged proton particles in the hydrogen nucleus to snap into alignment with the field. This happens because protons, and many other subatomic particles, have what's called a 'magnetic moment', which can be thought of as the magnetic field generated by the proton's positive charge as it moves.

Just as an ordinary magnet placed in an external magnetic field tends to align its north and south poles with the south and north poles, respectively, of the field (think of how a compass needle moves), so the magnetic moments of the protons are all made to swing into line by the magnetic field of the MRI scanner. The field strength needed to do this is enormous. The largest commercial scanners in operation today generate magnetic fields of up to 7 Tesla: 200,000 times the strength of Earth's natural magnetic field. The hefty magnets needed to generate such fields are the main reason why MRI scanners usually take up a whole room. It's also why no metal objects are allowed anywhere near an MRI scanner – even paperclips can become highly dangerous missiles when accelerated in such an intense field. Forget an MRI scan if you have a pacemaker fitted. And don't take magnetic media such as credit cards near one, as they will almost certainly be wiped.

Once all the hydrogen protons in the target area have been aligned by the magnetic field, they are bombarded with a pulse of radio waves. These waves are absorbed by some of the protons, knocking their magnetic moments out of alignment. A short time later the protons snap back in line with the field, emitting a pulse of radio waves back as they do so. This effect is called nuclear magnetic resonance (NMR), and was discovered in 1938 by US physicist Isidor Rabi, for which he won the Nobel Prize in Physics in 1944. Different kinds of tissue in the body – such as muscle and bone – have different magnetic properties, making them flip back at different rates. And this is what allows a picture of the body's internal structure to be built up.

Going functional

Functional magnetic resonance imaging (fMRI) was developed in the 1990s. It specializes in detecting changes in the levels of oxygenated blood present in certain areas of the brain. It was known in the late 19th century that increased brain activity raises blood oxygen levels in activated areas. Living tissue requires oxygen to convert its reserves of chemical energy into a usable form. The brain overcompensates for this demand by massively ramping up the blood flow to active areas, creating an oxygen surplus. This effect – known as 'blood-oxygen-level-dependence' (BOLD) – was discovered in 1990 by Japanese scientist Seiji Ogawa. FMRI works because oxygenated blood and deoxygenated blood have markedly different magnetic properties. When oxygenated, blood is said to be diamagnetic – meaning that it is repulsed by magnetic fields. Deoxygenated blood, on the other hand, is paramagnetic – it is the complete opposite, being attracted to magnetic fields. This makes the NMR radio signal from the two very different, and that's how active brain areas are spotted in an fMRI scan.

Being inside an fMRI machine can be a claustrophobic experience. The patient lies on a table that is then slid into a tube about 1 m (3 ft) wide and sometimes as little as 40 cm (16 in) high, slotting them deep into the heart of the scanner. The scanning process itself is extremely noisy – not unlike listening to a wrench in a spin dryer – and many patients have to wear ear plugs or headphones with music playing. Scans can last up to an hour, and during this time it's crucial the patient remains perfectly still. Movement by as little as a few millimetres can ruin the whole procedure. FMRI has revolutionized neurology, the branch of medical science dealing with the brain and nervous system. Now tumours can be imaged in detail, to resolutions of just a few millimetres, allowing doctors to plan brain surgery procedures with pin-point accuracy, maximizing their chances of success while minimizing the risk to the patient.

Tell me the truth

In recent years, the power of fMRI has been brought to bear on new tasks, often far removed from the sphere of medicine. Scientists believe that fMRI, and other brain scanning techniques, can be used to probe not only the health and wellbeing of our grey matter but also other aspects of brain function. They could reveal what our preferences are in terms of everything from fizzy drinks to politicians and even whether or not we're being economical with the truth. That's because each part of the brain serves a particular function. So, for example, a brain scan might show activity in the nucleus accumbens – a region of the brain associated with pleasure – when someone is happy. One study was able to tell the difference between feelings of love and lust: love was found to be associated with activity in the ventral tegmental area (which is responsible for manufacturing the pleasure hormone dopamine); lust is rooted instead in the amygdala, where many base emotions are processed, and the hippocampus, which is responsible for regulating feelings of thirst and hunger.

Liars can often be spotted by the sheer volume of activity in the brain. Whereas telling the truth simply involves recounting what's in your memory, fibbing is much harder work because your brain needs to constantly check each new part of the story against what's been said already to ensure consistency. And, sure enough, when test subjects were asked to lie deliberately while undergoing a brain scan, the scanner found a total of 14 brain regions active, compared to just seven when the subjects were being truthful. Some researchers have argued that this technology could help to improve the treatment of prisoners, eliminating the need for interrogation. However, some ethicists have voiced concerns that lie detector brain scans could be used to vindicate torture if the lie detector indicated the subject wasn't being truthful under conventional interrogation.

Most intriguing of all are some of the applications of this technology to marketing and PR. In 2004, for example, a study in the science journal *Nature* provided a fresh insight into the old Pepsi challenge TV commercials, where punters are asked to take a blindfold taste test of both Pepsi and its rival Coca-Cola. The researchers repeated the test, but this time they scanned the subjects' brains while they were sipping their drinks. About half of the blindfolded participants preferred the taste of Pepsi. Their brains showed activity in the ventrimedial cortex, an area connected with pleasure and reward. Then the experiment was repeated but without the blindfolds. Subjects were told which drink was which, and asked to make their choice again. This time, the number of people who said they preferred Pepsi dropped to about a quarter. Meanwhile in their brains, the prefrontal cortex lit up – an area thought to be responsible for some of our higher powers of reasoning. Brain scanning had revealed the power of branding and explained why Coca-Cola continues to beat Pepsi in the shops, even though people seem equally divided as to which tastes better.

Other experiments have been able to reveal people's preferences for different political parties. And scans have even shown up inherent feelings of racism in some test subjects when shown pictures of people from ethnic minority backgrounds. The scans revealed activity in brain areas linked to distrust. The subjects whose brains showed this behaviour had also scored highly in written tests designed to detect latent racism.

Dream watchers

If reading someone's thoughts seems incredible, one group of researchers even believe it may one day be possible to effectively download a person's memories directly from their brain. In 2008, US neurologist Jack Gallant and colleagues published a study in which he was able to tell which of a set of 120 different images a patient was looking at just from the pattern of activity that it established in their brain. He believes that it 'should soon be possible in principle to decode the visual content of mental processes like dreams, memory, and imagery'. If he's right, some bizarre possibilities may be feasible. For example, the memories locked away in the brain of a cooling corpse at a murder scene could be used to identify the killer.

Mind control

It is now possible to make devices that can be controlled purely through the power of your mind. The 'Emotiv Epoc' is a headset, principally designed for gamers, that can read the wearer's brainwaves. It translates them into electrical impulses that can then control a computer or game console. The idea is that if you think 'left' your on-screen avatar will move left, think 'open fire' and you will blaze away merrily with your weapon of choice. It can also measure emotional states such as frustration or excitement – and even facial expressions (by monitoring electrical activity in the face muscles) – to enhance the gaming experience.

The Epoc works using a suite of 16 electroencephalograph (EEG) sensors positioned over the skull to detect the wearer's brain waves. Brain waves are electromagnetic signals emitted from the brain because of currents that are generated as electrical charges are accelerated between brain cells – exactly the same mechanism by which a radio transmitter antenna works (see *How to cause a blackout*). Different thought processes require different impulses and so give off different patterns of waves.

The Epoc comes with a software peripheral called the 'EmoKey', which allows the user to map any thought signal detected by the headset to a keystroke shortcut. So, for example, if you want to be able to save your work without having to stop typing, you'd first think 'save' to produce a test thought pattern and then tell the EmoKey to save whatever you're working on whenever it detects this pattern in the future.

Emotiv believe the Epoc could be a massively empowering technology for the disabled, opening up the possibility of mind-controlled electric wheelchairs and household appliances. Perhaps more importantly it could give the most severely disabled patients access to computers and the internet, enabling them to socialize, shop and even do business through tools such as eBay. The device is on sale in the United States now, though it will require additional quality and safety certification before it can be released in the UK and Europe. Whether through EEG headsets or room-sized fMRI scanners, what's amazing is that human beings have learned the ability to read minds. Not through hypnosis or crystal balls – but thanks to breakthroughs in our fundamental understanding of physics.

GLOSSARY

anthropic principle
The fact that humans and other lifeforms exist places constraints on the laws of physics so that they permit the emergence of carbon-based life in our Universe. This is the anthropic principle.

antimatter
Every type of subatomic particle has an antimatter partner with the same mass and other properties, but opposite electric charge. When matter and antimatter particles meet they annihilate, converting their entire mass into energy.

asteroid
Chunks of rock left over from the birth of the Solar System that wander through space and occasionally collide with the Earth and other planets.

atoms and molecules
The smallest building blocks of the naturally occurring chemical elements are called atoms. Atoms can be bonded together to make molecules, the building blocks of more complex chemical compounds.

Big Bang
The event in which all the matter, energy, space and time of our Universe was created. It happened about 13.7 billion years ago.

black hole
An object with a gravitational field so strong that not even light can escape from it. Any object can form a black hole if it is squashed down to become dense enough.

chaos theory
The emergence of seemingly random behaviour from well-ordered physical systems. Chaos is produced by sensitivity to

initial conditions, causing initially similar states to diverge rapidly.

conservation laws
If a quantity in physics does not change with time then it is said to obey a conservation law. Energy and momentum are examples where this applies.

convection
Hot air rises; cool air sinks. This process is called convection. It crops up everywhere from a pan on a stove, to weather systems, to the insides of stars.

Copenhagen interpretation
An early view of quantum theory, where subatomic particles obey quantum laws until they are measured, at which point their wave-functions 'collapse' and they behave like classical objects.

dark matter and energy
Only about 5 per cent of the mass and energy in our Universe can be accounted for by the light it emits. The rest is invisible, and its presence is inferred purely by its gravitational effects. This is dark matter and dark energy.

eclipse
When the Moon passes directly between Earth and the Sun, it temporarily blots out the Sun's light. This effect is a solar eclipse. Lunar eclipses canalso occur when Earth passes between the Sun and the Moon.

electromagnetism
The theory of electromagnetism, formulated by 19th-century Scottish physicist James Clerk Maxwell, unified electricity and magnetism (the force that makes compass needles move) revealing them to be just different aspects of the same thing.

electroweak theory
In the 1970s, physicists were able to unify electromagnetism with the weak nuclear force. They called their model the electroweak theory.

energy
The capacity of a system to do work. It is measured in Joules (J).

One J is roughly the energy required to lift a 100 g mass by a distance of 1 m in Earth's gravity.

Fermi paradox

If there are spacefaring alien races elsewhere in the Universe then they should be here by now. But we don't see them. This is the Fermi paradox, which is used to argue against the existence of extraterrestrial intelligence.

fluid dynamics

Whereas Newton's laws of motion describe the dynamics of solid objects, fluid dynamics is concerned with the movement of gases and liquids under the action of forces.

general relativity

An extension of special relativity to include the force of gravity. It does this by bending the flat space and time of the special theory according to the matter and energy it contains.

grand unified theory

A theory of particle physics that pulls together the electroweak and strong nuclear forces into one entity. A 'theory of everything' includes gravity too.

gravitational lensing

Because general relativity represents gravity as curvature of space, it can bend the path of light beams too. This means that light from galaxies on the far side of the universe can get bent and magnified by the gravity of intervening matter.

Higgs boson

A fundamental particle of matter thought to have given all other particles in the Universe their mass. Researchers at the Large Hadron Collider are searching for evidence that it exists.

Hubble's law

The space of our Universe is expanding. This was discovered in the 1920s by American astronomer Edwin Hubble. Hubble's law states that the expansion rate increases with distance.

inertia

The resistance of a body to moving or changing its existing state of motion. To all intents and purposes inertia is the same thing as mass.

inflation

A colossal growth spurt, during which space is thought to have expanded by a factor of 10^{26} – a 1 followed by 26 zeroes – in a tiny fraction of a second shortly after the Big Bang.

Kepler's laws

Laws describing the motion of the planets worked out by German Johannes Kepler in 1605.

many worlds interpretation

An alternative to the Copenhagen interpretation of quantum theory, the many worlds interpretation ascribes the weird behaviour of quantum particles to interference between parallel universes.

Newton's laws of motion

Three simple laws of motion formulated in the 17th century by British scientist Sir Isaac Newton.

Newton's theory of gravity

A theory that predicts the gravitational force that two or more massive bodies exert on one another. Although superseded by general relativity, it still gives a very good approximation for weak fields.

nuclear fission

The process of obtaining energy by splitting heavy atomic nuclei in two. It is the basis of all nuclear power stations in use today.

nuclear fusion

Reactions that generate energy by joining together light atomic nuclei, used in modern nuclear weapons. Fusion is the power source of the Sun.

optics

The study of the passage of light rays. It explains phenomena such as reflection, refraction (the bending of light) and diffraction (how light is spread out by narrow apertures).

phase transition
A process by which matter undergoes a fundamental shift from one state to another. Examples include the boiling of water to steam, and the spontaneous breaking of particle physics symmetries in the early Universe.

photoelectric effect
The basis for solar power, explained by Albert Einstein, who showed how some metals generate electricity when ultraviolet light falls on them.

plate tectonics
Earth's crust is divided into a number of pieces that jostle together. Their motion is governed by a field known as plate tectonics.

principle of least action
Perhaps the single most important concept in physics, which says that physical systems always follow the path of least resistance.

quantum entanglement
A pair of 'entangled' quantum particles remain linked so that even when they are separated by a great distance, wiggling one will have an instantaneous effect on the other.

quantum theory
An often counterintuitive description of the world of subatomic particles. It says particles can sometimes behave like waves and, equally, waves sometimes act like particles.

radioactivity
The leaking of particles and radiation from the unstable nuclei of some atoms. Radioactivity takes three main forms: alpha particles, beta particles and gamma rays.

radio astronomy
The study of radio waves given off by far-away galaxies and cosmic gas clouds.

resonance
Like a tuning fork that's struck, all objects have a natural frequency.

When subjected to vibrations at this frequency they undergo violent oscillations. This effect is known as resonance.

Schrödinger's cat
A thought experiment based on the Copenhagen interpretation of quantum theory, which supposes that if a cat in a box lives or dies according to the behaviour of a quantum particle then the cat must be alive and dead at the same time.

simple harmonic motion
Oscillating movement, such as that of a pendulum, a guitar string or a mass bouncing on a spring.

special relativity
A theory formulated in 1905 by Albert Einstein, which rewrote Newton's laws of motion for objects moving at close to light speed. The theory predicted distortions in space and time as well as the famous equation $E=mc^2$.

spontaneous symmetry breaking
How unified theories of particle physics broke apart as the Universe cooled to give the theories seen today. For example, the electroweak theory split into the weak nuclear force and electromagnetism.

star
A vast ball of gas composed mainly of hydrogen, illuminated by nuclear fusion reactions in its core. The closest example is our own Sun.

statistical physics
The analysis of systems in physics composed of many particles. It works by applying the mathematical laws of statistics to give the system's average large-scale properties.

string theory
An attempt to remove some of the mathematical anomalies that arise from treating subatomic particles as zero-dimensional dots by treating them instead as one-dimensional 'strings'.

strong nuclear force
The force that glues together the quarks that protons and neutrons

are made of, and in turn binds these larger subatomic particles to form the nuclei of atoms.

subatomic particles
Matter is composed of atoms and molecules, which in turn are composed of subatomic particles, such as protons, neutrons and electrons.

superconductivity
Metals which, when cooled to within a smidgen of absolute zero temperature, lose all their electrical resistance. An electrical current in a loop of a superconductor will continue to circulate forever.

superfluid
Liquids that lose all their viscosity when cooled below a threshold temperature. An example is liquid helium.

supernova
A calamitous explosion that occurs when a massive star reaches the end of its life and throws off its outer layers. For a time, a supernova outshines the star's entire host galaxy.

thermodynamics
The physics of heat transfer and how that heat can be used to do useful work – for example, in engines.

uncertainty principle
It's one of the curiosities of quantum theory: the more accurately you measure one property of a subatomic particle, such as its position, the less accurately you are allowed to know other information about it, such as its speed. The principle was formulated in 1926 by German physicist Werner Heisenberg.

vacuum energy
Virtual particles mean empty space isn't as empty as you might think. The effect is called vacuum energy and the dark energy that dominates the Universe is one consequence of it.

virtual particles
Space is filled with a sea of virtual particles that can pop in and out of existence, so long as their energy and the time they exist for satisfy the uncertainty principle.

viscosity

The 'stickiness' of a liquid. Water has low viscosity while syrup has high viscosity. The theory of viscosity is essential for understanding the drag forces that act on cars, ships and aeroplanes.

wave theory

Drop a stone in a pond and ripples carry the energy of the impact outwards. Other forms of wave motion determine the physics of radiation and how sound travels from its source to your ear. Wave theory is the description of these processes.

weak nuclear force

One of two forces that operate within the nuclei of atoms. It is the force responsible for radioactive 'beta' decay.

work

In physicist's parlance, work is the force applied to an object multiplied by the distance the object moves in response. It is measured in Joules, the same units as energy.

INDEX